Jasta Colors

The Markings and Painting Schemes of the German Jagdstaffeln in World War I

Volume 2

Evolution and Background of the Markings
Summer 1914 – Winter 1916/1917

For Historians and Modelers

Bruno Schmäling
Jörn Leckscheid

Jasta Colors

The Markings and Painting Schemes of the German Jagdstaffeln in World War I

Volume 2

Evolution and Background of the Markings
Summer 1914 – Winter 1916/1917

For Historians and Modelers

Bruno Schmäling

Jörn Leckscheid

Interested in WWI aviation? Join The League of WWI Aviation Historians (www.overthefront.com), Cross & Cockade International (www.crossandcockade.com), and Das Propellerblatt (www.propellerblatt.de)

Design and layout: Jack Herris
Cover design: Aaron Weaver
Front cover painting: Russell Smith
Color profiles: Alexandr Kasakov
Digital photo editing: Aaron Weaver & Jack Herris

Aeronaut Books

Books for Enthusiasts by Enthusiasts
www.aeronautbooks.com

Publisher's Cataloging-in-Publication data

Schmäling, Bruno
    Jasta Colors Volume 2 / by Bruno Schmäling.
    p. cm.
    ISBN 978-1-953201-01-0
1. World War, 1914–1918 -- Aerial operations, German. 2. Fighter pilots -- Germany. 3. Aeronautics, Military -- Germany -- History. II. Title.

ND237 .S6322 2011
759.13 --dc22   2011904920

This series of books is dedicated with gratefulness to my late teacher, mentor, and friend

Alex Imrie

His research in the markings and painting of the German Jagdstaffels was ground-breaking for all research in this field to follow.

In memory of Nathalia Imrie

With a special thanks for the beautiful time to Eriká, Nathalia, and Alasdair Imrie.

The Authors

Bruno Schmäling

He dealt intensively with the history of the German military air service of the First World War for decades. From the 1970s until the first part of the 1980s he was able to interview 70 former members of the German Jagdstaffeln of the 1st World War and copied their photos and documents. For many years he had the privilege to cooperate closely with aviation historian Alex Imrie who become his friend and mentor. His other mentor was the German aviation historian Dr. Gustav Bock who also took him under his wings.

Jörn Leckscheid

Another student of the "Alex Imrie Academy" was again the indispensable co-worker in the creation of the book. From the first moment we started our cooperation I was, and am, impressed by Jörn's knowledge about the technical aspects of the airplanes, his expertise on the fabric finish of the aircraft, and his excellent evaluation and interpretation of the photos. Without his assistance the book and especially the color profiles could not be done in this quality.

Collaborators

A very special thanks to my "comrades in research" in the 1970s and 1980s:

Reinhard Behrend
late Dr. Gustav Bock
late Neal O'Connor
late Peter M. Grosz
late William R. Puglisi
Hans-Eberhardt Krüger
Gero von Langsdorff

late Paul Leaman
late Heinz J. Nowarra
late Herbert Schulz
late Wolfgang Schulz
Michael Schmeelke
late Ludwig Zacharias

My tribute goes to the forgotten German aviation historians of the 1920s and 1930s. Without their basic work, the research today would be hardly possible:

Ludwig Himmelstoß
Hanns Isemann
Egon Krüger
Theodor Melcher

Kurt Schorler
Fritz Schmidt
Willi Stiansny
Erich Tornuß

For support on this series of books my special thanks also go to:

Rainer Absmeier
Richard Alexander (formerly of Wingnut Wings)
Winfried Bock
Thomas Genth
Jon Gutman
Trevor Henshaw
Jack Herris
Reinhard Kastner
Andrew King (Old Rhinebeck Aerodrome)
Dr. Volker Koos
Stephen Lawson
Jim Miller
Piotr Mrozowski (Aviation Museum Kraków)

David Méchin
Colin Owers
Lt. Col. Terry "Taz" Philips, U.S.A.F. ret.
Tom Polapink (Old Rhinebeck Aerodrome)
Johan Ryheul
Josef Scott
Igor Starikov
Marton Szigeti
Dr. Hannes Täger
Aaron Weaver
Tobias Weber
Reinhard Zankl

A very special thanks to my friends and colleagues:

Greg VanWyngarden who over decades conducted dedicated research about the history of the German Jagdstaffeln and the painting of the aircraft, who generously supported my work with the results of his research.

Lance Bronnenkant who, over years, has been an excellent colleague in research and supported me with biographic data for the book.

Without their support this series of books could not be made in this quality.

Unless otherwise noted, all photographs originate with the collection of Bruno Schmäling.

Acknowledgements

My heartfelt gratitude is extended to the former German fighter pilots and members of the groundcrew of the First World War who generously supported my research with their memories, photos, and documents in the years 1974–1984. Sitting together with these great gentlemen and listening to their stories is a precious memory, and the foundation and motivation to write my books. Even today, when I work on my books, the memory comes back and takes me back to the time when I sat face to face with theses fascinating gentlemen and listened to their experiences.

My special thanks go also to the over 150 relatives of the former pilots who also generously supported me with their photos and documents.

Moreover, my special thanks go to Prof. Dr. Gisela Röper, Ludwig Maximilian Universität of Munich for her support in the 1970s to shape a concept for successful interviews with World War I pilots.

Last but very far from least I would like to thank my then girlfriend and now since many years my wife Kerstin from the bottom of my heart for all her understanding, patience, and support. She accompanied me on a number of visits to former fighter pilots and their relatives and supported me in documenting these meetings. Without her understanding, support, and patience it would not be possible to build up such an extensive hobby with about 18,000 photos and the associated thousands of documents!

Bruno Schmäling

We give our very special thanks to Director Frau Dr. Haggenmüller and the staff of the Bavarian Main State Archive, Dept. 4 for the continued support of our research over all those years.

The authors are also very thankful to Mr. Piotr Mrozowski of the Aviation Museum in Krakow. He supported us in a generous way with the results of his extraordinary research literally covering the fabric finish of German two-seater planes in 1914–1916. We also enjoyed the cooperation with the members of the "Interessengemeinschaft Luftfahrt 1900–1920" in Germany.

We are fortunate that artist Russel Smith did the beautiful cover painting again for this book; it is a pleasure to work with such a prolific artist! With Alexsandr Kazakov we are happy to cooperate with a new artist who masterfully provided the excellent color profiles for this book. The work of both artists illustrates the colorful appearance of German fighter planes in a highly realistic manner that truly breathes life into the pages of this book.

As with my previous books, Aaron Weaver created the cover. He also did most of the beautiful colorizations of the black and white photos, which are of great importance to create an impression on how the colorful German fighter planes really looked like in the air. Jim Miller was so kind to give us the permission to use one of his colorized photos for the book.

Of course, our special thanks goes to the best publisher in the world, our friend Jack Herris, as well as to our colleagues Hannes Täger, Greg VanWyngarden, Lance Bronnenkant, and Reinhard Kastner who kindly reviewed the manuscript.

Bruno Schmäling & Jörn Leckscheid

The authors and publisher of this book series would be very pleased if modelers would send photos of their scale models made according to the information (color profiles) in this book series.
We want to give modelers a chance that photos of their models will be published in a separate publication.
Please send photos to: **jherris@me.com**

Table of Contents

Preface

The first book in this series portrayed examples to illustrate which source material is available to ensure an authentic representation of the Staffel identifications and personal insignia of German World War One fighter aircraft.

In this second book of the series, we first describe the development of aircraft identifications and paint schemes covering the timespan from the summer of 1914 to the winter of 1916/17 in the context of the military development of the German Fliegertruppe. The two were inextricably linked and the development of aircraft markings and livery must be seen in this context.

The first flying units of the German Empire had the task of observing the enemy and, if the opportunity arose, also dropping some bombs. With the beginning of the positional warfare the additional task of directing artillery fire on enemy positions arose. This led to a specialization of the units, and in addition to the previous Feldflieger-Abteilungen (Field-aviation units), Kampfgeschwader der Obersten Heeresleitung (Fighting Squadron of the Army High Command) and Artillerieflieger-Abteilungen (Artillery-aviation units) were created.

As early as the beginning of the war, the first individual aircraft crews began to apply personal identification markings to their aircraft, thus initiating a development that was increasingly adopted by other aircraft crews as well.

The markings and paint schemes of these individual early aircraft largely served as the role model for the aircraft of the later to be formed German Jagdstaffeln, the first of which were created in the summer of 1916. For this reason, I present the markings and paint schemes of the two-seater aircraft of these units during 1914 – 1916, on the basis of individual examples. Without this exemplary presentation the history of the painting schemes that were later applied to the single-seater fighter planes would be incomplete.

The book focuses on the formation of the first ten German Jagdstaffeln as well as the individual markings and paint schemes of the aircraft operated by these units. The timespan from the late summer 1916 to the winter of 1916/17 is covered by illustrating a number of their aircraft. In this context, the military reasons for the colorful identifications and paint schemes of the aircraft are also discussed.

As in Volume I, the pilots or crews of the aircraft shown are introduced with a brief curriculum vitae, as far as this is possible on the basis of available documents.

And also as in the previous volume, the primary source of information originates in the notes the interviews and correspondence with 120 former members of the German Jagdstaffel conducted by Alex Imrie in the 1950s and 1960s, Dr. Gustav Bock between 1960 and 1970, and Michael Schmeelke and me in the 1970s through the 1980s. The recollections, photographs, and records of these former Staffel members are indispensable to this series of books. During the meetings, our interlocutors often described their experiences in detail and in vivid colors, and I have adopted one or the other experience for the book in order to bring the atmosphere in the flying units back to life.

As in Volume I, I could not resist to give a little insight into my meetings with former fighter pilots or their relatives. The meetings with them were always the highlight of my research and I have many fond memories of them, which should not be lost. The personal contacts brought life into the research and allowed a picture to emerge about the events of that time. The motivation for my books is based on these memories and the desire not to let the history of these fascinating people be forgotten.

Another basis of this series of book are the documents, notes, and information of the first generation of German aviation historians who gathered important information in the 1920s and 1930s. Alex Imrie had the privilege of working with some of them, such as Willy Stiasny, Ludwig Himmelstoß, and Fritz Schmidt. The American aviation historian William "Bill" Puglisi worked with Erich Tornuß and purchased the collection of Theodor Melcher, as well as fragments of the largely lost collection of Kurt Schorler. Egon Krüger assisted Peter M. Grosz in his research.

Dr. Gustav Bock worked very closely with Erich Tornuß for many years. It was Erich Tornuß who, in the 1930s, transcribed excerpts from the war diaries of the German Jagdstaffeln archived in the Reich Archives in Potsdam. Dr. Bock revised, supplemented, and corrected these documents and information based on his extensive research in various German archives, an achievement whose importance can hardly be overestimated. At that time, the work of amateur historians was uncommon, and this group of German aviation enthusiasts had to overcome a number of difficulties before they could gain access to the state archives.

Erich Tornuß had also numerous personal and written contacts with former German fighter pilots in the 1930s. In the process, he also gathered a number of very interesting pieces of information about the painting of the aircraft. I requested Dr. Bock to also ask Erich Tornuß about any information he may have found out concerning the paint schemes of the Jagdstaffel aircraft and pilot's personal markings during their meetings and ongoing correspondence. In the transcripts he made for Bill Puglisi and Dr. Gustav Bock, he had previously included this information only sporadically, since Dr. Bock and Bill Puglisi were mainly interested in the historical events. Due to my request Dr. Bock started to query Erich Tornuß about this information specifically and compiled it for me. Finally, he brought me in direct contact with Erich Tornuß. This turned out to be a stroke of good luck for me for my research!

I personally had the pleasure of working with Herrn Ludwig Zacharias and Herrn Egon Krüger and received extremely interesting information and documentation from them.

Another special stroke of luck was the contact with Herrn Herbert Schulz from Hamburg. He had already started as a young man to collect information about the history of the Jagdstaffeln 2 and 5. Through the Pour le Mérite aviator Paul Bäumer he came into contact with the "Kameradschaftliche Vereinigung der Jasta Boelcke" (Comradeship Association of Jasta Boelcke) and was even able to attend a few meetings. His records were of particular value for this book.

Further interesting information came from Herrn Gero von Langsdorff, the son of the pilot, aircraft constructor and aviation historian Werner von Langsdorff, who, among other things, published several books about WW I airmen also under the pseudonym "Thor Goote". The records he left behind proves that he was in contact with a number of former fighter pilots in the 1930s. For example, his book "Rangehen ist alles," published in novel form, is based on interviews with former members of Jagdstaffeln 2 and 11.[1]

The achievement of this generation of aviation historians cannot be overestimated. They were in contact with many former fighter pilots at a time when their experiences of World War I were just a few years old. They had access to the original records still available at the time in the Reichsarchiv in Potsdam, as well as private records. Since there were no copying machines at that time, they had no choice but to painstakingly transcribe the documents by hand. Their records are, due to the many lost original documents, of inestimable value.

Few of these pioneers had photographic equipment at their disposal to photograph the original photos of the airmen with whom they were in contact. They had to hope to get copies of photos as gifts. Thus, their photo collections were often not very large, but that in no way diminishes the value of their work.

Unfortunately, hardly any of this first generation of German aviation historians ever published anything. Considering the knowledge they had at their disposal, one can barely measure this loss for the documentation of the history of the German Fliegertruppe in World War I.

The book is also based on official and semi-official documents of the German Fliegertruppe of the 1st World War. Official documents on the allocation of fighter planes to the Jagdstaffeln in the autumn of 1916 were found in the Bavarian Main State Archives, Dept. 4, in Munich. These are documents that are invaluable for this book.

Another indispensable source are thousands of photos and many records, such as flight books, documents, letters, and written memories, which had been made available to me by relatives of former fighter pilots and other aviation historians. Finally, I used various publications that were published after World War I, in the 1920s, such as books, newspaper articles, etc.

Historical research is always the composition of different fragments that have to be put together like in a puzzle. In contrast to a jigsaw puzzle, however, parts are always missing, and the gaps left by this fact must be filled by extrapolation and interpretation in order to obtain an overall picture. (2) For this reason, I always indicate in the text the sources of the photos, documents, records, and statements on which the representation of the color profiles is based.

Unfortunately, some publications about the Fliegertruppe of the First World War lack such source references! These publications are simply useless for the presentation of the topic because it is not possible to determine which content is based on historical sources and what is left to the interpretation or even fantasy of the author.

Over the course of 40 years of working on this subject, I have experienced that sometimes the discovery of a single new photo is sufficient to reveal completely new aspects concerning the identification and painting of the aircraft of a Staffel. Historical representation is therefore always to be regarded as a dynamic process. Accordingly, misinterpretations are always possible due to incomplete original sources. For this reason, Jörn and I are grateful for additional source material, comments, and corrections. They will then be integrated into one of the following volumes of the series.

The authors connect the work on this book also with the hope that it motivates other aviation historians to further research on this topic and that modelers build beautiful models of German military aircraft or fighters with the help of this book.

Bruno Schmäling, Summer 2022

Introduction

The fighter planes operated by the German Jagdstaffeln were the most colorful war planes which ever participated in a war. These bright markings did not suddenly appear out of nowhere. The application of colorful painting schemes to the aircraft already started at the beginning of the war and was connected to the development of the German Fliegertruppe and the way this troop was commanded.

The young men who climbed into the cockpits of the early military airplanes were fully aware that their lives depended upon the flying machines made of wood, steel tube and fabric, a method of construction that was, by modern standards, anything but trustworthy. As a consequence, a special relationship between the human being and the machine quickly began to develop. The desire to personalize the aircraft quickly began to arise, and for this reason it was only logical that several of these aircraft were soon given names or nicknames or were marked otherwise by their aircrews. During the few short years of civilian aviation before the war some individual markings had already been applied to airframes, but this was only the case on a few single, isolated aircraft.

Soon after the outbreak of the war individual markings and paint schemes began to appear on the aircraft of the Feldflieger-Abteilungen and with the Kampfgeschwader der Obersten Heeresleitung. The markings which were applied by these early units would later, when the first Jagdstaffeln were formed in the summer of 1916, serve as the role model for the painting schemes that were applied to the single-seater aircraft of the fighter units. For this reason, it is essential to turn an eye towards the evolution of the markings and paint schemes applied to German military airplanes from the outbreak of the war in the late summer 1914 up to the middle of 1916, without making any claim that this book is a comprehensive study in this field.

During the 1st World War no official system which was intended to identify the different units or aircraft existed in the German Fliegertruppe. All aircraft were to be marked with the "Eisernes Kreuz" (Iron Cross) as a national marking, in different variations. From spring of 1918 onwards, the Balkenkreuz (literarily translated "bar cross" - the terms "Greek Cross" or "Maltese-Cross" – previously used in several publications – are unknown in Germany) was introduced and it was applied in different variations until the end of the war.

Besides the marking(s) identifying the unit to which an aircraft belonged, every additional application of paint to an aircraft was left up to the pilot or crew of the airplane, pending, of course, the commander of the unit voiced no objections.

Stepping into the subject, it is necessary to differentiate between:

Staffel-markings:
Homogenous marking or color that identified the airplane as belonging to a certain unit (Staffel, Fliegerabteilung or Geschwader).

Personal markings:
Color(s) or emblem(s) that identified the personal airplane of a pilot or a crew.

Fabric Finish of the aircraft
Additionally, we attempt to present the fabric finish of the aircraft as realistically as possible, based on the intensive research of Mr. Piotr Mrozowski of the Aviation Museum in Krakow and Jörn Leckscheid who also did a long and intensive research in this field.

Prologue

Airfield of a German Jagdstaffel in Northern France in October 1916

The Staffelführer of a German Jagdstaffel, an Oberleutnant, gazed thoughtfully into the early autumn skies of northern France. It had been just over a month since his Jagdstaffel had been newly formed, one of the first ever. He gazed toward the west and his eyes searched the sky. Another officer, also of the rank of Oberleutnant, stepped up beside him. Earlier he had served as an artilleryman, as could be seen from the black band of his service cap, which sat slightly askew on his head.

The Staffelführer looked to him questioningly. *"Anything new?"*

The former Artilleryman, who was assigned to the Jagdstaffel as officer for special duties (z.b.V.), shook his head, *"No, so far no one has reported the landing or crash of an Albatros D II."*

"We're missing two machines," the Staffelführer noted, adding, *"I hope each of them came down somewhere in one piece."*

"I have spoken to the Flakgruppen-command and little Müller continues to phone all the other units. As soon as he learns anything of importance, he will inform us."

The Staffelführer took his walking stick, turned from the wood of a broken propeller, and poked around on the ground as if he would find a clue to the missing planes there. Then he looked up, *"Once again we failed to stay together in formation. When we reached the English squadron, three of our planes were missing. Then, when I gave the signal to attack, the planes flying farther back attacked too late and before we knew it, the British had formed a defensive ring. As a result, we flew right into the middle of their hail of bullets."*

"But everyone attacked?"

"Yes, of course, the guys don't lack grit! But the opportunity to break up the formation was lost. It's not the lack of courage or the will to go for it. It's still the lack of coordination in skilled Staffel flying during the attack."

"But all of our airmen are very eager to do so," the z. b. V. officer interjected proudly.

The Staffelführer nodded. *"Of course, you're right, of course they're full of zeal, but if we don't manage to stay together as a formation, then..."*

They both fell silent and went back to searching the sky.

Finally, the Staffelführer turned again to his z. b. V.: *"We succeeded in pushing off two "Vickers" and brought one down. (1) Suddenly a group of the* small English lattice tails appeared and attacked us from above. They were the same planes as the one the Boelcke Staffel had brought down intact some time ago. *Then things got really nasty! The English fighters are not as good as our Albatros, but still, they are rather unpleasant opponents. In the middle of the thickest hustle and bustle, another German Jagdstaffel joined in and then it went haywire."*

He paused his speech, reached into his breast pocket where his pocket watch was and looked at the dial. Then he shook his head. *"The last drop of fuel has finally been used up now. There's no point in waiting here any longer, let's go to the office."*

Both officers turned and walked slowly across the grass runway to the wooden building that served as the office. The office shack stood next to the large tents that housed the sleek Albatros D II fighters.

The Staffelführer turned back to his z. b. V.:

"To continue the story, one of the little "lattice tails" also went down, but the "Lords" managed to escort the remaining "Vickers" safely across the lines. But of course, that was not the end of the story. I sought to collect my flock. Five I found immediately. To pick up the others, I had to fly a larger circle several times. I was able to collect the rest, but there was no trace of two planes. I have not seen them since the attack on the big "Vickers". Since no more Englishmen appeared at the front, we flew back. Fortunately, the three crates we had first lost had already landed."

The z.b.V. officer nodded, *"Yes, they had lost sight of the Staffel and had turned back."*

"It was the best they could do, partly because our two bunnies were there."

Both officers had almost reached the wooden hut, when an orderly, it was the Gefr. Müller, because of his height only called the little Müller, stormed out of the door. He greeted them briefly, then announced with a joyful look: *"Our two airmen have landed with a Jagdstaffel to the north. I just got off the phone with one of them. They didn't find our planes after the dogfight and then joined the other Jagdstaffel. After refueling, they will fly back. "*

The Staffelführer smiled with relief, *"Thank God, they're back safely."*

"Then let's go to the mess hall," suggested the z. b. V. officer.

The Staffelführer nodded. *"Did you report our aerial victory?"*

"Yes."

Above: The pilots of the newly formed Jagdstaffel 10 in front of the start-house, October 1916: from left: 1st Uffz. Barth, 2nd Vzfw. Heldmann, 3rd Offz.Stellv. Viereck, 4th Lt. Hess, 5th Lt. Dr. med. Weber, 6th Lt. Bordfeld, 7th (unknown), 8th Lt. Nernst, 9th Oblt. Volkmann, 10th Offz. Stellv. Hebben.

Left: Lt. Godt of Jagdstaffel 6 poses proudly in the pilot's seat of his Albatros D I. The holes caused by enemy bullets were patched with cockades, a common method in the entire Fliegertruppe. (G. VanWyngarden)

Above: The Pilots of Jagdstaffel 2 at Lagnicourt airfield. In the background an Albatros D I of the Staffel.

"And what about eyewitnesses from the ground?"

"The officer of an anti-aircraft unit has confirmed to me the crash of a Vickers. He intends to send me a report tomorrow. "

"That's good."

"Yes, but the other Jagdstaffel also lays claim to it. "

"That's what I thought, I'm curious to see who gets it. They fly the same Albatros as we do. From the ground, the planes are indistinguishable. "

"Which makes confirmation difficult."

"Yes, but that's not the main problem. In the air, I can hardly distinguish which aircraft is in which position. If one or more aircraft are lost, I can't immediately tell who is missing. We would have to mark the crates somehow so that folks can recognize each other in the air. "

The Staffelführer paused, changed direction, and walked toward two trestled-up Albatros D IIs standing in front of the nearest tent. Shortly thereafter, both officers stood in front of an Albatros on which three mechanics were busily working.

"Everything all right?" the Staffelführer asked one of the men in the oil-stained outfits.

He shook his head. *"Herr Oberleutnant, on this machine here, we're replacing a radiator right now. It took two hits. Then on both machines we still have to check and repair the bullet holes on the fuselage. "*

"Are there multiple machines damaged?"

"Almost everyone took hits Herr Oberleutnant."

The Staffelführer nodded. *"Thank you, carry on."*

He turned back to his z. b. V.: *"This is the result if the gentlemen from the other side manage to form a defensive ring. I wonder if all the machines will be operational tomorrow?"*

"You know our mechanics are doing their best." One heard that the officer z.b.V., as head of the non-flying personnel, was proud of his "black men".

The Staffelführer smiled and patted his z. b. V. on the shoulder. *"I know that I know that."*

Then he takes a step toward one of the Albatros D IIs and stands next to the fuselage. He pointed to the area behind the pilots's seat: *"The solution would be to put a badge here on the fuselage, so we can recognize each other, and I can see who is in which position in the formation, who has problems or who is missing."*

"And what identifier are you thinking of?"

"When I flew with the Kagohl, we had big numbers painted on our crates. That made it easy for us to maintain formation and notice when one was missing."[2]

"Then we should do it the same way."

"I think that would be the best solution. Each plane gets a big number, on either side of the fuselage."

The Staffelführer stroked one hand across the top

14

Below: A typical view of an airfield operated by a German Jagdstaffel in the late summer 1916. On the left is the wooden hut with the office, on the right the tents to accommodate the aircraft. The squadron members also have a motorcycle at their disposal. (R. Zankl)

of the smooth, honey-colored fuselage. *"I think the number should be on the top of the fuselage, too. If we fly staggered upward, you can see the number from the top as well. They need to be big and noticeable."*

"All right, I'll take care of the paint. What color do you think we should use?"

"A color that is highly visible, maybe red?"

The Staffelführer nodded to the z. b. V. *"We'll do that. I'm counting on you. You're a natural at organizing, so see what you can scrounge up."*

The z. b. V. officer pretended to assume a military posture, brought his hand casually to the brim of his service cap, and said with a grin, *"At your command, Herr Staffelführer."*

"Then let's wait in the casino for our two lost bunnies. Afterwards, I want to see the whole bunch in the casino. We need to discuss the last mission again."

The very next day, the officer z. b. V. was on his way to the Armee-Flugpark (Army Aviation Supply depot) to look for suitable paint stock there. After he failed to obtain the desired red paint there, he had

his driver take him to the Army depot. He searched the depot and after some time he found what he was looking for. The red paint he found would be sufficient to apply the new identification numbers to all the machines flown by the Staffel. Satisfied, the z. b. V. officer had the paper bags with the paint pigments and several buckets of cellulose for mixing the paint stowed in the car. Then he ordered his driver to take the road back to the aerodrome.

When he arrived at the airfield, he had the paint unloaded and taken to one of the airplane tents that had been set up.

Then he beckoned to the Werkmeister (foreman), a stocky man with a Kaiser Wilhelm mustache: *"We'll put a number on both sides and on the top of each plane. Can you do that? I've made a list here of who should get which number. "*

The Werkmeister in the rank of an Vizefeldwebel circled an Albatros D II standing in the tent. He nodded. *"Yes, of course. We should cut templates out of thick cardboard to do this, so it looks neat and uniform."*

Above: Albatros D II of an unidentified Jagdstaffel marked with the number "3" as personal identification. (G. VanWyngarden)

"A good idea," praised the z. b. V. officer, adding, *"Can't Schmidt do it, he's a sign painter by trade and seems to be rather talented?"*

"Of course, Herr Oberleutnant, I'll talk to Schmidt and then we'll get the cardboard to cut out the numbers."

When all the aircraft were on the airfield in the late afternoon of the following day, they were pushed into the tents and the mechanics started painting. The very next morning, a large red number shone on both sides and on the top of the fuselage of each of the sleek Albatros D IIs. There were the numbers 1 - 9 and then the letters A, B, C.

Satisfied, the Staffelführer nodded. Now he would immediately recognize who was flying where and who might be missing. In addition, he could keep a better eye on the two bunnies. These were two young Leutnants who had been transferred to the Jagdstaffel from a Feldflieger-Abteilung unit a few days ago.

Preliminary Note

The **first three chapters give a** brief overview of the development of the German Fliegertruppe in the years 1914–1916. For each year number of examples portraying the identification and painting markings of various aircraft of the Fliegereinheiten (flying units) are presented.

The descriptions do not claim to offer a complete picture of the identifications of this early period. This would go far beyond the scope of this book. It would be desirable, however, if other aviation historians would devote themselves to this subject in detail at some point.

1. The German Fliegertruppe in 1914

At the conclusion of mobilization, the German Army had 33 Feldflieger-Abteilungen (Field Aviation units), 8 Festungsflieger-Abteilungen (Fortress Flying units), and 8 Etappenflugparks (Aeroplan supply depot) at their disposal.[1] A total of 232 aircraft were available to the German Fliegertruppe at this time.[2]

Each of the eight Armee-Ober-Kommandos (A.O.K.) and each general command of the 26 active army corps was assigned a Feldflieger-Abteilung that was intended to conduct daytime reconnaissance over enemy territory. The reserve army corps did not have a Feldflieger-Abteilung at their disposal. Soon after the beginning of hostilities, most of the Festungsflieger-Abteilungen were converted into Feldflieger-Abteilungen and dispatched to the front[3].

The crews of the Feldflieger-Abteilungen had only just been able to hint at their performance capabilities during the peacetime maneuvers and were eagerly awaiting to conduct their first frontline missions. At last, the time had come to prove the military value of their weapon. The crews took off as soon as the planes were ready for action, sometimes without having received an order to embark on a mission from the Army Corps to which they were assigned. A short time later, they were able to present accurate reports of the enemy's movements to the baffled staffs. Sometimes their information was so precise that they were not believed at first until other sources confirmed their information.

Ernst von Hoeppner, the later commanding general of the Luftstreitkräfte wrote in his book „*Deutschlands Krieg in der Luft 1914–1915*": "*The brilliant reconnaissance results of the aviators led to a complete turnaround in the evaluation of the new weapon after the first weeks of the war. What had not been believed in peacetime had come to pass: The aviators had completely supplanted the cavalry as a long-range reconnaissance tool. Thanks to their excellent performance, they enjoyed an esteem that lifted them out of the secondary role conceived in peace and placed them on an equal footing with the main weapons.*"[4]

The achievements of the German Feldflieger-Abteilungen in 1914 is all the more astonishing because the German Fliegertruppe initially had no tight organization at the front to regulate its operations. The Feldflieger-Abteilungen were therefore largely forced to organize their missions themselves.

It was not until September/October 1914 that the first German armies on the Western Front gradually created the position of the Stabsoffizier der Flieger (staff officer for aviation within an arms), which at least provided a point of contact for the deployment of air units within an army. As early as 30 Sept 1914,

Above: German airplanes on the Western Front. Two L.V.G. "B" of Bavarian Feldflieger-Abteilung 5 in the park at Bondues Castle north of Lille, in the area of the German 6th Army, in the fall of 1914 (R. Kastner).

the Army High Command of the German 2nd Army established the position of staff officer for aviation, followed by the German 5th Army in October 1914.[5] However, as the name implies, this was a staff position that had only an advisory function to the Army High Command (AOK), but no direct command authority.

1.1 Examples of the Individual Markings of Aircraft – 1914

The respect that the aircrews had gained within a short time led to the fact that individual crews began to paint their aircraft individually:

Feldflieger-Abteilung 16

The Abteilung was mobile on 1 August 1914 and was under the command of Hptm. Schmoeger. It was deployed with the German 8th Army on the Eastern Front.[1]

Feldflieger-Abteilung 16, L.V.G. B, late summer/autumn 1914, Profiles 1 & 1a (cutout face)

The fuselage and wings of the L.V.G. B were covered with yellowish linen fabric covered with a protective dope. The L.V.G. B of Feldflieger-Abteilung 16 was marked with a face sticking out its tongue at the enemy as a personal insignia on the underside of the engine cowling. This may have been one of the earliest personal paint schemes on a German military aircraft serving at the front.

Feldflieger-Abteilung 23

The unit was one of the 33 Feldflieger-Abteilungen mobilized at the start of the war on 1 August 1914. The Abteilung was deployed in the area of the German 2nd Army. The Abteilungsführer was Oblt. Vogel von Falkenstein.[2]

The painting of the aircraft is documented by the photo album of Lt. Wilhelm Pier, from the Marton Szigeti collection.

Above: Mechanics and guards of L.V.G. B of Feldflieger-Abteilung 16 in September 1914 on the Eastern Front. The aircraft has been marked with a personal identification of a face sticking out its tongue at the enemy. (R. Zankl)

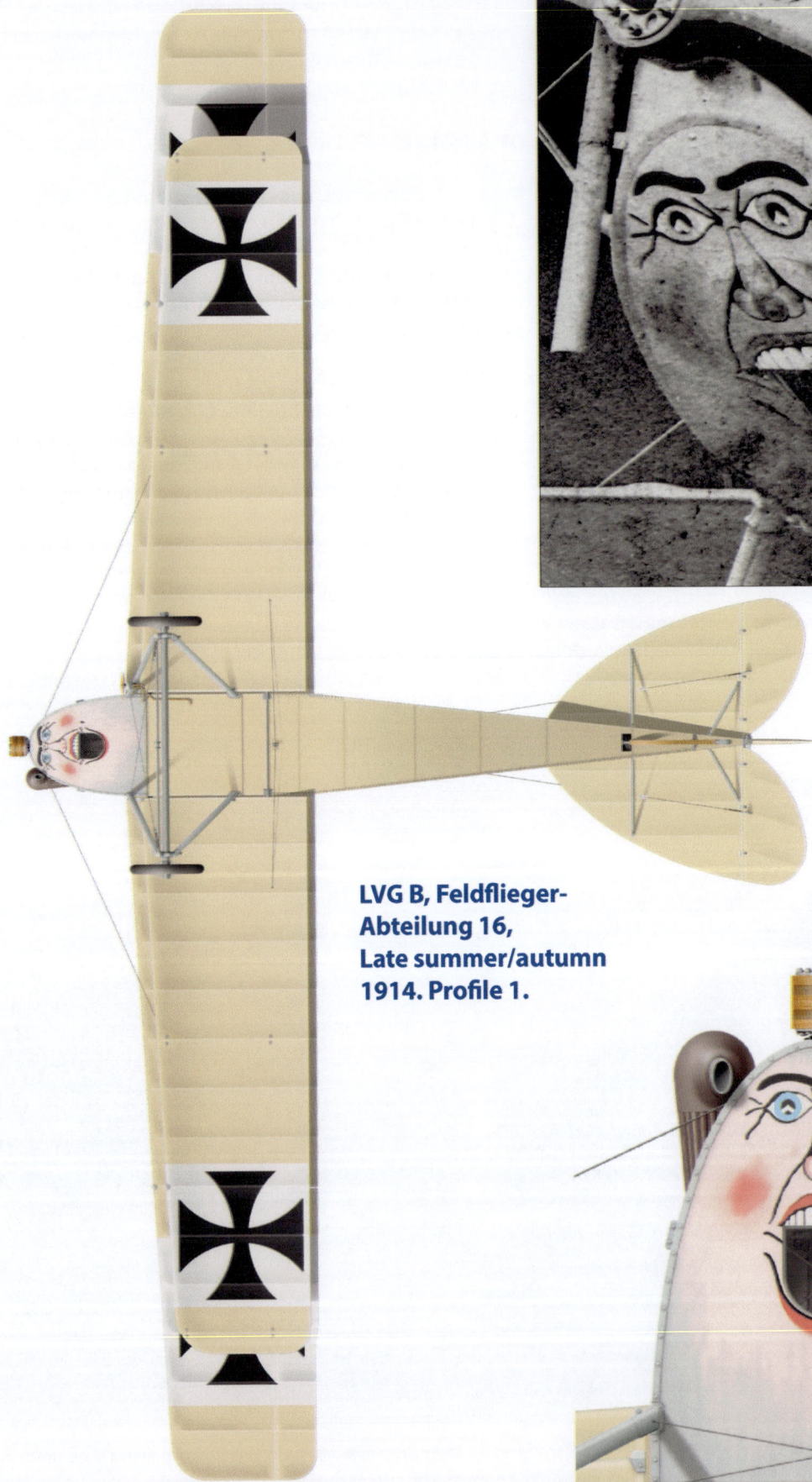

Above: Blow-up of the face of L.V.G. B of Feldflieger-Abteilung 16 (R. Zankl).

LVG B, Feldflieger-Abteilung 16, Late summer/autumn 1914. Profile 1.

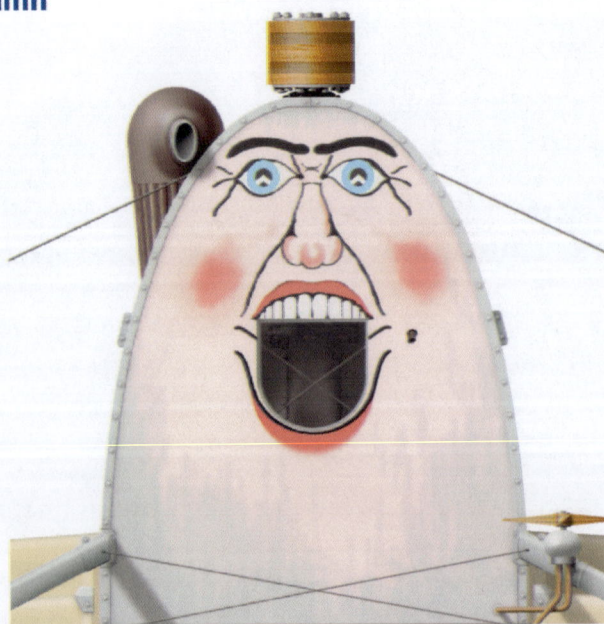

Fertig zum Aufstieg.

Above: A contemporary color drawing of an L.V.G. B I (Otto) of Bavarian Feldflieger-Abteilung 3 with the title: "Fertig zum Aufstieg" ("Ready to take off") recreates the atmosphere in the Flieger-Abteilungen at the beginning of the war, as well as portraying the beige fabric covering of the aircraft. (R. Zankl)

D.F.W. B II Feldflieger-Abteilung 23, October 1914. Profile 2.

Feldflieger-Abteilung 23 D. F. W. B II, October 1914, Profile 2

The early D.F.W. B II had a yellowish plywood fuselage covered with a translucent protective varnish. The wings were covered with a yellowish linen fabric, which was also covered with a protective clear dope. This is also documented by a color drawing of the D.F.W. B I B.451/14 captured by the French.[3]

The crew of the D.F.W. of Feldflieger-Abteilung 23 had a victor's wreath consisting of laurel and oak leaves with a black, white, and red ribbon painted on the rear part of the fuselage. **As the laurel wreath is only partially visible, we have used a contemporary**

Above: A D.F.W. B II of Feldflieger-Abteilung 23 in October 1914. Listed neatly on the fuselage are the frontline missions for the period September–October 1914. To the left of it is the top of the painted victor's wreath of laurel and oak leaves can be seen. (M. Szigeti)

illustration to serve as a pattern for the color profile. Painted in front of it was a chronological mission listing of their frontline flights, presumably in black and red, applied to the fuselage side:

«6.9.14 Romilly-Nogents-Orbais
7.9.14 Fere-Champenoise-Sesanne-Orbais
9.9.14 Montmirail-La Ferbe- Orbais
12.9.14 - 21.9.14 Rhyme Isles
25.9.14 Reims-Epernay-Dormans-Bazancourt
26.9.14 Reims-Les Petites-Epernay-Fismes-
    Bazancourt
3.10.14 Mourmelon-Epernay-Ville en Tardenois-
    Bazancourt»[4]

The crews were proud of their missions at the front and such "mission reports" can also be found in the photos of other German frontline aircraft taken at the time.

Above: The contemporary postcard illustration of a German victor's wreath which was used as a template for the colour profile.

Feldflieger-Abteilung 22/ Brieftauben-Abteilung Ostende

Feldflieger-Abteilung 22 was also one of the aviation units that mobilized on August 1, 1914. The **Abteilungsführer at that** time was Hptm. von Blomberg.[5] The Fokker A II/M.5L was initially in

Above: Captured German aircraft displayed in Paris in 1914. Visible in the foreground is D.F.W. B I B.451/14, a "Taube" can be seen in the upper right corner. The "Iron Cross" national insignia are also applied to the upper surfaces of the lower wing. (G. VanWyngarden)

Contemporary color drawing of D.F.W. B I. B.451/14 shows the yellowish linen fabric covering the aircraft. (G. VanWyngarden)

Below: Lt. Otto Parschau, Brieftauben-Abteilung-Ostende (B.A.O.), with one of his mechanics in front of his green Fokker A II/M.5L, which was initially flown by Oblt. Waldemar von Buttlar.

service with this unit and was then transferred to the Brieftauben-Abteilung Ostende (BAO). This unit had been formed on 27 November 1914, with the purpose of conducting strategic bombing. On 20 December 1915, the BAO was converted into the Kampfgeschwader der Obersten Heeresleitung (Kagohl) I.[6]

The Fokker A II/M.5L first flown by Oblt. Waldemar von Buttlar and later by Lt. Otto Parschau also had a mission report painted onto the fuselage side. It is striking that on the only existing photo of the aircraft no national insignia can be seen on the wings.

Otto Parschau wrote about **this aircraft in his** letter to Anthony Fokker dated 25 May 1915:

"Something that will interest you and which should not happen every day is that now the green bird that Oblt. von Buttlar flew in Belgium was flown by me from September 28 to November 30, 14 in Champagne, from December 1, 14 - February 2, 15 in Flanders, from February 3, 15 - February 20, 15 in Alsace-Lorraine, from March 6,(15) - April 20, 15 in East Prussia from April 24, 15 - ? in Western Galicia. The other day Lt. von Oertzen said to me that the plane should be in the museum. Right he is."[7]

Based on the letter, we know that the plane was painted green. The reason might have been that Oblt. Waldemar von Buttlar was a member of the "Kurhessisches Jägerbataillon Nr. 11" which had a green uniform like all "Jäger" (**hunter**). However, the letter does not give any information about what green tone was used to paint the plane or what color was used to mark the mission list.

Black lettering on a dark green fuselage should have been hard to read, same as red lettering. We therefore suspect that the fuselage was painted in a light green color that had faded by the elements over its long period of use of over five months. In that case, black or red lettering would be good enough to see. Since the inscription on the fuselage side recording the sorties flown was more of a personal record of the pilots Waldemar von Buttlar and Otto Parschau, rather than serving as a recognition marking in the air, black lettering is shown in the illustration of the aircraft.

Likewise, it is unclear if the wings were also green. **We opted** for the light green variant with black lettering and wings also painted light green. This could be the reason that no national insignia is **visible on the top side.** Since what appears to be a white tip of the rudder is just visible at left, the color plate shows a white rudder with an Iron Cross, as seen on other contemporary examples of the type.

The aircraft may have been one of the first whose personal paint scheme was a reference to the pilot's peacetime uniform. The display of the uniform color or the color of the cap bands was later adopted by

Oblt. Waldemar von Buttlar/ Lt. Otto Parschau, Fokker A II/M.5L, Feldflieger-Abteilung 22/ Brieftauben-Abteilung Ostende, Autumn 1914 – Spring 1915, Profiles 3 and 3a (top view).

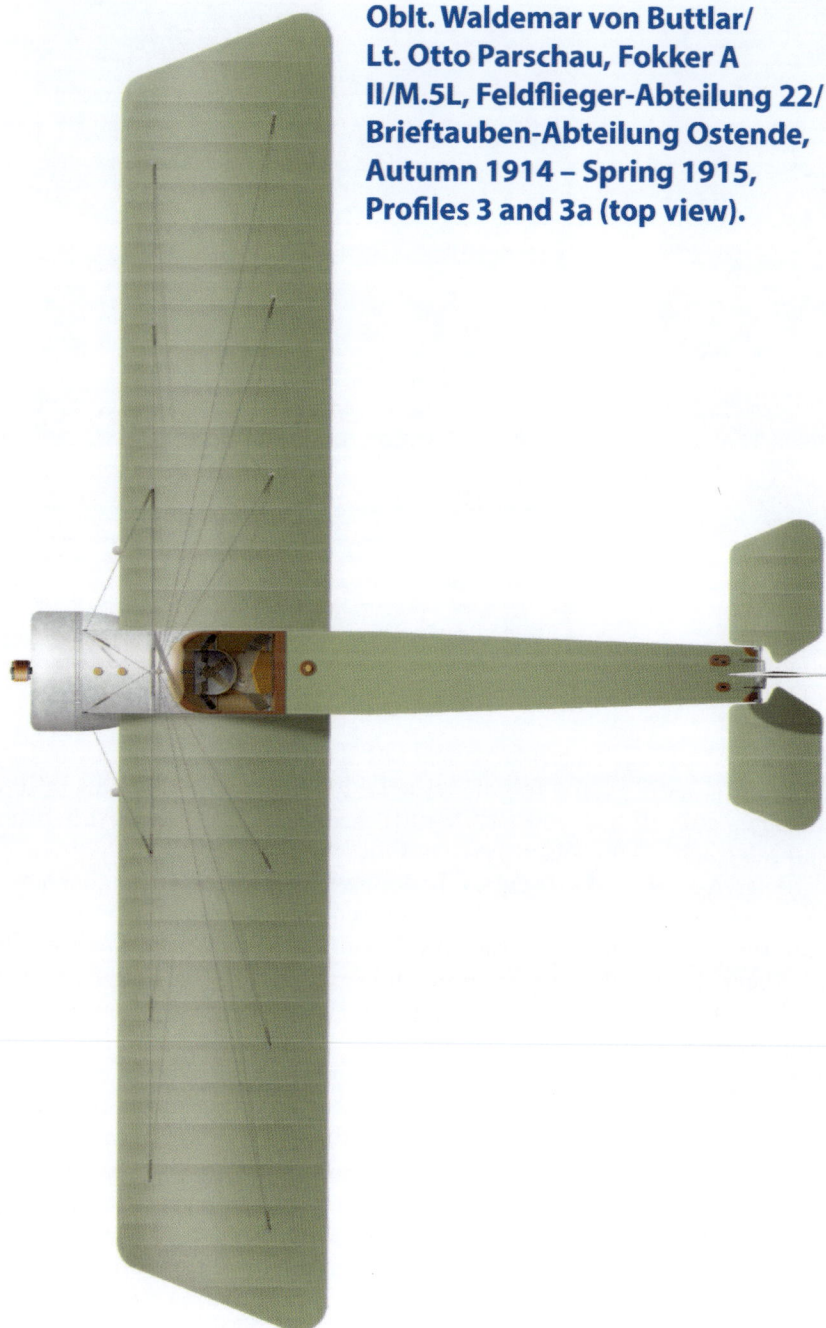

a larger number of fighter pilots as the personal color on their aircraft. Even if one belonged to the Fliegertruppe, pilots usually felt deeply attached to their parent units, and this was especially true for officers.

Waldemar Walrab Freiherr von Buttlar was born in St. Quirin on 26 March 1885. He acquired pilot's license No. 167 as early as March 1912 and took part in the South Germany Flight in 1912, as well as in the Prinz Heinrich Flights in 1913 and 1914. Together with Lt. von Schröder he took third place with the L.V.G. Apparatus No. 12 in **May** 1914. In 1914, he was a pilot with Feldflieger-Abteilung 22. All that is known about his later life is that he later worked for Bayerische Motorenwerke (BMW). He passed away on 18 April 1952.[8]

Otto Parschau was born on 11 November 1890 in Klutznitz near Allenstein in East Prussia. After graduating from high school in the autumn of 1910, he joined Infantry Regiment No. 151 as an officer candidate and was promoted to lieutenant a year later after completing the officer training course at the Hanover

Above: Lt. . Otto Parschau as a member of the B.A.O. with his shepherd dog. In the background the railroad train, which served the members of the B.A.O. as quarters and casino. Thus, the unit could be quickly relocated.

Right: The peacetime uniform of the "Kurhessisches Jägerbataillon Nr. 11" which very likely served as the model for the green paint scheme of the Fokker A II/M.5L. In the case of the aircraft the green color looks lighter and faded after long frontline service.

War College. In March 1913 he transferred to the Fliegertruppe, which was being formed, and from April 1 was trained as an aircraft pilot at the Johannisthal Fliegerschule. On 4 July, after passing his pilot's exam, he received pilot's license No. 455 and was able to make a name for himself by winning various flying competitions until the outbreak of war the following summer.

He belonged to Feldflieger-Abteilung 22 from August 1914. On 1 December 1914, he was transferred to Brieftauben-Abteilung Ostende.[9] According to various sources, he also belonged to Feldflieger-Abteilung 42 in between. However, this

seems rather unlikely due to the fact that this unit was not mobile until 5 December 1914. At this time, he was already serving with the BAO.[10]

Otto Parschau achieved his 1st aerial victory on 11 October 1914 and scored his second kill shortly before Christmas. In February 1916, he joined the Staffel on the Verdun front, and here increased the number of his aerial victories to four by the end of March, for which he was named in the German Army Report on 22 March 1916. In June he transferred to Kagohl 1, which operated on the Somme, scored three kills there within four days, and for this was awarded the Knight's Cross of the Royal House Order of Hohenzollern on 3 July 1916. Transferred to Feldflieger-Abteilung 32 five days later, he achieved his 8th aerial victory on 9 July 1916, whereupon he became the fifth fighter pilot to be awarded the Order Pour le Mérite the

Above: Oblt. Waldemar von Buttlar, left, and Lt. Otto Parschau, right, in front of Fokker E I 1/15. The photo was taken on the premises of the Fokker factory in Schwerin-Görries. (T. Genth)

Above: Ready for action. An Aviatik B of Feldflieger-Abteilung 34 marked with the number "6" in the winter of 1914/1915 (R. Zankl).

Above: Aviatik B 344/14 of Feldflieger-Abteilung 34 ended up on its nose following an imperfect landing in the winter 1914/1915. The number "3" applied to the fuselage fabric indicates that the use of numbers for identification purposes started early in the war. (R. Zankl)

Aviatik B I, Feldflieger-Abteilung 34, late autumn 1914 / winter 1914–1915, Profiles 4 and 4a.

The color painting of an Aviatik B I used at the front in 1914 was created in 1915 at the Flieger-Ersatz-Abteilung 10 in Böblingen, where the subject of the painting was used as a training aircraft. The Aviatik B I carries a personal identification in the colors of the German Empire, black-white-red, on the fuselage. Behind it an A.G.O. C I marked with the black and white identification of Armee-Abteilung Gaede has also made it into the painting. (G. VanWyngarden)

following day. Appointed leader of Kampfeinsitzer-Kommandos Nord, which was attached to Feldflieger-Abteilung 32, on 14 July, he was severely wounded by an abdominal bullet in aerial combat with several enemy aircraft only a week later, on 21 July 1916, and died the same night in the field hospital at Grevillers.[11]

Feldflieger-Abteilung 34

The order to form Feldflieger-Abteilung 34 was issued on 22 August 1914, and the division was mobile and moved to the front on 27 September 1914 [12]

The painting of the aircraft is documented by postcards and the corresponding notes in the Reinhard Zankl collection.

Aviatik B I, Feldflieger-Abteilung 34, late autumn 1914 / winter 1914-1915, Profiles 4 and 4a.

The Aviatik B had a linen-covered fuselage and wings covered with a protective dope that gave the aircraft a yellowish appearance.[13]

Available photos prove that the aircraft operated by Feldflieger-Abteilung 34 were marked with an identification system consisting of the numbers 1 to 6. This may have been one of the first systematic identifications of the aircraft operated by a

Feldflieger-Abteilung. The aircraft described has number 6 painted on the fuselage. The color of the number is not known and was interpreted as red for better visibility but could possibly have been black. For this reason, we show both variations. The photos with the identifiers are from postcards sent by the Abteilung in March 1915. However, it is very likely that the identifications of the aircraft had already been introduced in late autumn 1914 and were still in use in spring 1915.[14]

1.2 Summary

According to an analysis of all available photographs, the vast majority of German aircraft in service at the front in 1914 had no individual paint scheme or identification. The number of their own and enemy aircraft was relatively small; moreover, the crews of the field aviation divisions flew their missions largely alone. Therefore, a special identification was not yet necessary. However, one had to reckon with being fired upon by one's own ground troops. For this reason, the national insignia was sometimes also applied to the lower surfaces of the upper wing, and sometimes even to the bottom of the fuselage. A number of aircraft also carried the Iron Cross on the top of the lower wing. This marking served to provide recognition by other German aircraft flying at a higher altitude.

2. The German Fliegertruppe in 1915

During the first two months of the war, to the surprise of the **army headquarters, the German Feldflieger-Abteilungen had fulfilled their task perfectly** and had become the eye of the troops. This situation began to change when, in October 1914, the first Allied aircraft armed with air-cooled machine guns began to appear in the skies of the Western Front.[1] The German aircrews, armed only with carbines, had little effective equipment available to ward off this new threat. From the winter of 1914/15 at the latest, Allied aircrews dominated the skies over the Western Front. If a suitably armed Allied aircraft appeared in the sky, the German crews had no choice but to take flight if they did not want to be helplessly shot down. The fact that in a few rare cases an Allied plane could be shot down by a chance hit or the outstanding performance of a German crew did not change this.

In Germany, only water-cooled machine guns were available at that time, which were not suitable for installation in aircraft because of their weight. As a first countermeasure from the German side, aircrews obtained captured Allied air-cooled machine guns and armed their aircraft with these. This was a difficult undertaking, as the number of these looted items was relatively small. A number of Feldflieger-Abteilungen sent their special-duties officers out to obtain captured machine guns and even offered cash awards for obtaining them. If a Hotchkiss machine gun could finally be found after much effort, its use was limited. After all, the corresponding Allied ammunition had to be acquired as well.

By the spring of 1915, the previous organizational and technical failures and deficiencies had become so obvious that on 11 March 1915, the position of the Chef des Feldflugwesens **(Chief of Field Aviation)** was created. He became the head of the Army's entire aviation and airship system, and the air defense. The Army Weather Service were also placed under his command. Lieutenant Colonel Herrmann Thomsen was appointed to take up this position, which was often abbreviated "Feldflugchef". Major Wilhelm Siegert, who had previously made a name for himself as an advisor on aviation-related matters to the Supreme Army Command, became Thomsen's staff officer. At the same time, the flying units were detached from the **Verkehrstruppe (transport troops) to which they had hitherto belonged. The Etappen-Flugparks were detached from the rear areas, given the new designation of Armeeflugparks, and placed under the command of the Stabsoffizier der Flieger (staff officer for aviation within an army). The Inspektion der Fliegertruppe (inspectorate of military**

Below: Even in the summer of 1915, unarmed aircraft were still in service with the Feldflieger-Abteilungen. One example is L.V.G. B II 1048/15 of Feldflieger-Abteilung 2. Lt. Heinz von Beaulieu-Marconnay flew a number of missions as an observer with this aircraft. He was the brother of the later Pour le Mérite holder Lt. Olivier von Beaulieu-Marconnay.

Above: On 11 March 1915, Oberstleutnant Herrmann Thomsen was appointed Feldflugchef (Chief of Field Aviation). Under his command, the successful reorganization of the German Fliegertruppe was orchestrated. (L. Bronnekant)

Above: Major Wilhelm Siegert in his office. He was instrumental in creating the position of Feldflugchec and served as its Stabschef (Chief of Staff).

Above: The Stabsoffizier der Flieger of the German 5th Army, Hptm. Wilhelm Haehnelt, right, in conversation with aircraft designer Anthony Fokker. .

aviation) was also now under the command of the Feldflugchef. The position of the Stabsoffizier der Flieger was created at each Armee-Ober-Kommando [2]

These measures finally led to a tighter and much more effective organization of the Fliegertruppe. Likewise, training and the provision of supplies improved markedly. The aircraft and engine industry could now plan and produce according to the requirements from the front. The reorganization had a very rapid effect. The situation at the front improved markedly when the first "C"-type aircraft, armed with air-cooled machine guns, also arrived at the front in the spring of 1915. In the meantime, the Parabellum MG 14, and a version of the MG 08 adapted for use in aircraft, had become available in sufficient numbers. In the first C planes, the observer still took up the front seat, as was common in the early B types, which made handling the

Above: Aviatik C I of Bavarian Feldflieger-Abteilung 8 with Lt. Erwin Wenig in the pilot's seat. The observer sat in front of the pilot in this type of aircraft. In this case the crew mounted a second Parabellum machine gun to the aircraft. This made sure that the observer did not have to lift the machine gun from one side to the other and adjust it on the rail in case of aerial combat.

machine gun difficult. He had to be careful as hell not to hit the wing struts by mistake. The disadvantage of this positioning of the aircrew was quickly recognized, and the seats were swapped in the types that followed. Now the pilot sat in front, while behind him the observer found sufficient space for his activities and had a far better field of fire.

In May 1915, the German Fliegertruppe consisted of the following units:

72 Feldflieger-Abteilungen, 2 Festungsflieger-Abteilungen, 1 Fliegerkorps der Obersten Heeresleitung, 18 Armee-Flugparks at the front, and 11 Flieger-Ersatz-Abteilungen in Germany.[3].

In the meantime, however, the air combat situation had once again developed to the disadvantage of the German Fliegertruppe. From April 1915, the first French Morane-Saulnier L fighters appeared at the front, with the machine gun able to fire forward through the propeller. This was made possible by deflectors mounted on the back of the propeller blades. These deflectors deflected projectiles as soon as they hit the propeller. **The German aircrews could defend themselves, if their aircraft was armed with a machine gun, but their lumbering two-seaters were clearly inferior to the fast and maneuverable French single- seaters.**

The first French aerial victory with the Morane-Saulnier L was achieved by the well-known pre-war pilot Roland Garros, who shot down a German aircraft on 1 April 1915. He won two more aerial victories on 15 April and 18 April 1915, until he was forced to make an emergency landing on the German side after his third victory. **Despite the fact that he set fire to his aircraft, the deflectors mounted to the rear surfaces of the propeller blades were discovered.** The Feldflugchef immediately ordered to undertake trials of a copy of the deflector system, replacing the Hotchkiss MG with a German Parabellum LMG 14. The tests showed that the deflectors had withstood the French copper bullets but were penetrated by the German chrome-nickel steel jacket bullets.

Shortly after Garros' deflection system was

Above: The French pilot Roland Garros in front of his Morane Sauliner L which featured metal deflectors mounted on the rear of the propeller blades. These were intended to prevent damage to the propeller when firing with the machine gun. (J. Herris)

Below: Lt. Kurt Wintgens in the pilot's seat of his Fokker E I 5/15 with Bavarian Feldflieger-Abteilung 6 at Saarburg airfield in July 1915. He was one of the first successful German single-seat-fighter pilots. Note the use of the tripod-mounted headrest.

32

Above: A Pfalz E II (probably 225/15 or 228/15) of Bavarian Feldflieger-Abteilung 8 at Colmar-Nord airfield in Alsace. A small rear-view mirror has been fitted on the left side of the cockpit.

inspected, Anthony Fokker, a Dutch aircraft designer working in Germany, informed the Inspektion der **Fliegertruppe** that an MG synchronization device could be inspected at his company that precluded damage to the propellers. Fokker, however, not only had a working MG synchronizer, but his M5K light monoplane was also an aircraft suitable for the installation of this system. The type was given the designation M5K (MG) after the installation of a Parabellum LMG 14. The Fokker E series was quickly developed from this design.

With the introduction of the Fokker single-seaters, the tide turned back in favor of the German Fliegertruppe in the summer of 1915. The Fokker monoplanes, which were soon joined by similar Pfalz monoplanes, became the bane of the Allied flying units on the Western Front with their fixed machine gun firing through the propeller arc. This was achieved despite the fact that the operational capabilities of the Fokker fighter monoplanes were limited due to their small numbers. In addition, the pilots of the single-seater fighters were strictly forbidden to cross the front, as the German side feared that the synchronization mechanism might fall into Allied hands. Despite this restriction, the use of these aircraft led to a complete turnaround in the air combat situation and a period began that went down in Allied air war history as the "Fokker Scourge."

These aircraft were fundamental in the establishment of the history of the German fighter weapon. Associated with them began the careers of

the first successful German fighter pilots such as Oswald Boelcke, Max Immelmann, Otto Parschau, Hans-Joachim Buddecke, Kurt Wintgens and Walter Höhndorf.

The original plan was to equip each Feldflieger-Abteilung with a Fokker single-seater fighter. Initially, the Feldflieger-Abteilungen operating in front sectors with higher enemy aerial activity were to be equipped with the single-**seaters first**. After only a few months, the allocation of individual Fokkers to all Feldflieger-Abteilungen engaged in frontline operations proved to be impractical, since enemy resistance was increasing in volume. In addition, considerable problems of forwarding news of observation of enemy aerial activity from the front became apparent, which often prevented rapid focal deployment of several Fokkers on the same target. In order to achieve effective coordination of the defense, as early as the autumn of 1915, individual single-seater fighters began to be grouped with Feldflieger-Abteilungen in main combat areas. These **units received several** Fokker or Pfalz fighter monoplanes. In order to achieve this goal, other Feldflieger-Abteilungen were obliged to transfer their Fokkers or were not assigned any at all.

This consolidation resulted in the first **Kampfeinsitzer-Kommandos (single-seater fighter detachment) at the end of 1915, beginning** of 1916, in which most Fokker or Pfalz fighter aircraft of an army were combined. The Kampfeinsitzer-Kommandos were not a unitized unit and therefore no uniform designation was assigned to them.

Thus, **depending on the army,** designations such as Fokkerstaffeln, Fokker-Kommandos, Abwehr-Kommandos, A.O.K-Staffel or Kampfeinsitzer-Kommandos appear in various official documents. In many cases, the airmen remained subordinate to their previous Feldflieger-Abteilung and were merely commanded to the single-seater unit. Since this command could be revoked at any time, turnover of personnel was considerable, especially in the early days. This situation did not improve until the command was combined with a transfer to the respective Feldflieger-Abteilung to which the single-seaters were attached.

By October 1915, the equipping of all Fliegerabteilungen on the Western Front with "C" aircraft armed with machine guns had been more or less completed. In order to direct artillery fire, nine Artillerieflieger-Abteilungen (artillery aviation units) were distributed among six armies on the Western Front. Similarly, a larger number of Fokker fighter monoplanes, which by now had been designated "E-planes", were distributed among the armies. In addition, each army possessed a few large twin-engine aircraft, so-called "G-planes."[4]

2.1 Examples of the Individual Markings of Aircraft in 1915

Above: A dog-face was marked on the engine cowling of this Pfalz Parasol operated by Bavarian Feldflieger-Abteilung 7. On the fuselage side, just below the pilot's seat, a small "picture frame" bordered with green oak leaves has been painted. It is possible that the name of the crew was later marked here. (R. Kastner)

Bavarian Feldflieger-Abteilung 7

The Abteilung was formed on 15 January 1915 at the Bavarian Flieger-Ersatz-Abteilung and was considered mobile from that day on.[1] The Bavarian Feldflieger-Abteilung 7 was located in the spring of 1915 at the airfield Hazavant Ferme, a farm in the area of the Armee-Abteilung (army detachment) of Strantz.[2] **The** equipment of the Abteilung consisted, among other things, of L.V.G. (Otto) B Is. Light Pfalz Parasol aircraft were also in use.

Above: Uffz. Gustav Diekmannshemke and his basset hound, who may have served as the inspiration for the marking, take a seat on the engine cowling of the Pfalz Parasol. The national insignia is also visible on the elevators. (R. Kastner)

The markings of the aircraft of the unit in the spring of 1915 are evidenced by the photo albums of a former member of the Abteilung, which are in the possession of Reinhard Kastner.

Vzfw. Gustav Diekmannshemke Pfalz Parasol, Feldflieger-Abteilung 7b, Spring 1915 , Profiles 5 and 5a (front)

One of the most striking early personal paintings on a German military aircraft had **this Pfalz Parasol of** the Bavarian Feldflieger-Abteilung 7. On the

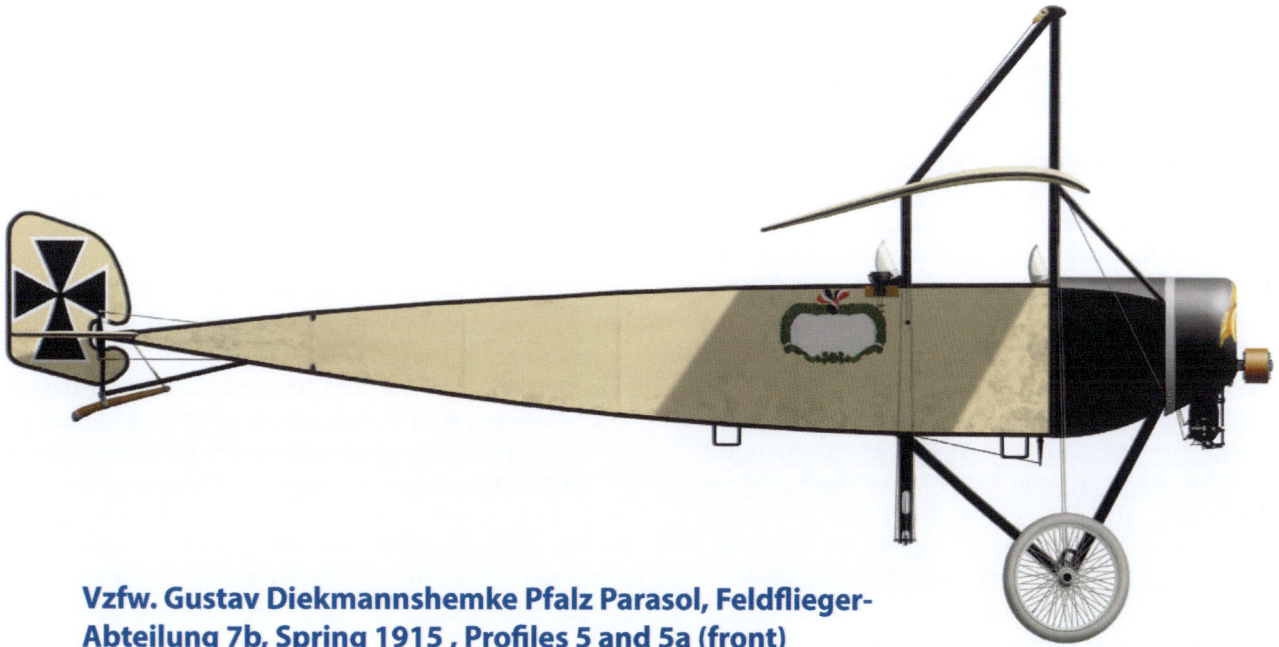

Vzfw. Gustav Diekmannshemke Pfalz Parasol, Feldflieger-Abteilung 7b, Spring 1915 , Profiles 5 and 5a (front)

engine cowling was painted the dangerous looking head of a dog. Painted on the fuselage side just below the cockpit was a white shield bordered with – presumably – **green** laurel. Likely the names of the crew were later to be applied there.[3] The Pfalz Parasol had a yellowish fabric covering with black edges. The engine cowling was also black. The aircraft was flown primarily by Vzfw. Gustav Diekmannshemke, and significant signs of use were apparent on the airframe when the photographs were taken.

Gustav Diekmannshemke was born on 14 October 1889 in Dorstfeld in West Prussia, now incorporated into the city of Dortmund. However, he enlisted in the Bavarian Army and went into the field with the Bavarian 5th Chevauleger Regiment. As early as 1913, he enlisted in the Fliegertruppe. He was trained on the Parasol at the Speyer-based Pfalz Flugzeugwerke in October 1914 and on the new monoplane at Fokker in Schwerin in November 1914. He saw frontline service as a pilot with Feldflieger-Abteilungen 3b, 7b and 19, and as a fighter pilot with Kampfeinsitzer Abteilung 1 and Kampfeinsitzer Abteilung "Großes-Hauptquartier". On 2 August 1916, he was transferred to the Bavarian Flieger-Ersatz-Abteilung at Schleißheim. Until the end of the war, he served as a flight instructor at the Bavarian Military Flying School 2.[4]

Feldflieger-Abteilung 53

The Abteilung was formed at Flieger-Ersatz-Abteilung 1 on 16 January 1915 and was mobile from 15 February 1915.[5]

The depicted identification and painting of the aircraft of the unit are based on the photo album of the Pour le Mérite-decorated ace Emil Thuy, who, after the war, fatally crashed near Smolensk on 11 June 1930. In the early 1980s, I was able to get in touch with Herrn Hans-Joachim Thuy, the son of the Pour le Mérite recipient. When I met him in the lobby of a hotel, I was immediately struck by the great resemblance to his father, and I could very well imagine what his **father had looked** like at a comparable age. Hans-Joachim Thuy was still very young when his father died, but he was able to recall what he had learned from his mother about what his father had told her about his time as a fighter pilot. To my great delight, Herr Thuy possessed two photo albums and a number of written documents, including reports of his father's missions, which he was happy to make available to me. Copies of some photos from this album I later provided to the American historian Ed Ferko, from where these photos eventually migrated to the library of the University of Texas, Dallas, from where they found use in some publications. Further documents and information about the history of Feldflieger-Abteilung 53 I received from the family of Kurt Freiherr von Crailsheim, whom I also visited around this time.

Lt. Kurt Freiherr von Crailsheim, Aircraft Pilot, Albatros C I, Feldflieger-Abteilung 53, Autumn 1915, Profile 6

The aircraft had a slanted band with the colors black-yellow-black around the fuselage, bordered on top with a white band. The colors black-yellow-

Below: Albatros C I of Feldflieger-Abteilung 53 in the fall of 1915, flown by Lt. Kurt Freiherr von Crailsheim. A fuselage band in the colors of the coat of arms of the von Crailsheim family, black-yellow-black, has been applied. A white band has also been added behind this marking.

Lt. Kurt Freiherr von Crailsheim, Aircraft Pilot, Albatros C I, Feldflieger-Abteilung 53, Autumn 1915, Profile 6

Above: Magnification of the multi-colored band of Kurt von Crailsheim´s aircraft: black-yellow-black with the white band.

black are, according to the family, the heraldic colors of the von Crailsheim family. Thus, this painting is an early example of aristocratic pilots using their coat of arms colors as a personal insignia on the aircraft. Ornately painted family crests sometimes also appeared on individual aircraft. The Albatros C I may have had a painted yellowish appearing plywood fuselage. The wings were "clear doped linen."

Kurt Freiherr von Crailsheim was born on 27 July 1892 in Ansbach in the Kingdom of Bavaria. He joined the Royal Württemberg Army and the infantry regiment Kaiser Wilhelm, King of Prussia (2nd Württemberg) No. 120. After being wounded, he enlisted in the Fliegertruppe and began his training on 1 December 1914, with Flieger-Ersatz-Abteilung 6 in Großenhain. After completing his training, he

Above: The coat of arms of the von Crailsheim family

was transferred to Feldflieger-Abteilung 53, stationed at Monthois, via the Army Air Corps of the German 3rd Army. After a series of successful missions, he was assigned to the Kampfeinsitzer-Abteilung in Mannheim in August 1915 for training as a "fighter pilot." After his return to Feldflieger-Abteilung 53, he flew his missions primarily with the Fokker fighter single-seater. On 22 September 1915, he achieved his first aerial victory flying his Fokker E II. Unfortunately, on 30 December, he crashed his Fokker E III on a landing approach and was so badly injured that he died five days later in the military hospital.[6]

Above: An A.G.O. C I of Bavarian Feldflieger-Abteilung 9 over the river Rhine plain. It still lacks the black and white markings of Armee-Abteilung Gaede. The silhouette of this type of aircraft could easily be mistaken for similar French aircraft from the ground.

Special aircraft markings of the Armee-Abteilung Gaede, early 1915 - autumn 1915

In 1915, the Armee-Abteilung Gaede (Army Department Gaede), whose front line extended from the Swiss border in the south to about the height of Strasbourg in the north, used a special identification marking in the form of black and white bands. The reason for this was that the AGO, Otto and Pfalz aircraft used by the army could easily be confused with similar-looking French aircraft types. This fact caused problems especially for air defense on the ground. For this reason, orders were issued in early 1915 to provide all Army Department aircraft with black and white bands. Since there were no exact regulations about the design of the identification, different, often quite creative implementations of these identifications developed. **The identification of the aircraft with the black and white bands of the Army Department Gaede is documented by original documents in the Bavarian Main State Archives in Munich and likewise, by a postcard from Lt. Walter Kiliani, observer of the Bavarian Feldflieger-**Abteilung 9.[7] In the autumn of 1915, the order to use these identifications was rescinded by Feldflugchef Oberstleutnant Hermann Thomsen.[8]

Bavarian Feldflieger-Abteilung 9b

The Abteilung was formed on 21 January 1915 at the Bavarian Flieger-Ersatz-Abteilung 1 and was mobile on 1 June 1915.[9] It was stationed at Colmar-Nord airfield in September 1915.[10] The identification and painting of the aircraft are well documented by the photo albums of former **Staffel members** Hans Auer, Erwin Wenig and Robert Dycke, as well as by the photo album of Oskar Müller, a member of Artillerieflieger-Abteilung 206, which was also stationed there.

Vzfw. Georg Pfleiderer, Flugzeugführer, AEG C I, Feldflieger-Abteilung 9b, Armee-Abteilung-Gaede, Summer 1915, Profile 7

The aircraft had a black cat arching its back ready for fight as a personal insignia, in addition to the black and white identification of the Gaede Army Department. **The fabric covering of the A.E.G. C I is shown in light blue because former airmen referred to these planes as the: "the light blue A.E.G. two-seaters".**

Georg Pfleiderer was born on 11 May 1892 in

Right: .The color drawing of Lt. Walter Kiliani, observer of the Bavarian Feldflieger-Abteilung 9 shows him with his AGO C I in aerial combat with a French aircraft. The caption reads: Bayerische Fliegerabteilung 9b. Aerial fight over the Vosges. (R. Kastner)

Deutscher „Ago" Doppeldecker
der
Bayr. Flieger Abtlg. N:9 Colmar.

Sehr ähnlich und leicht zu verwechseln mit einem
französischen Voisindoppeldecker.

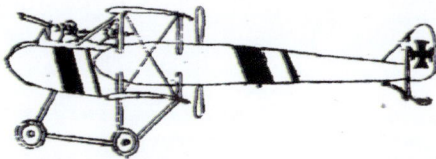

Farbe:
Weißlich-gelb.

In der Luft durchschimmernd.

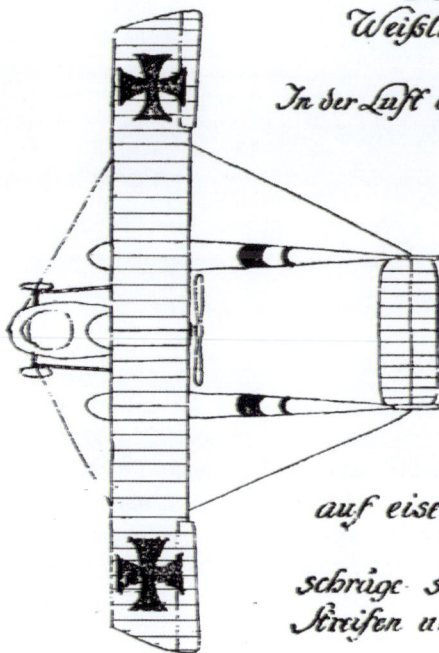

Achtung
auf eiserne Kreuze
und
schräge schwarzweiße
Streifen um die Rümpfe.

FIEGERABT. No 9

Left: Information leaflet prepared for the German anti-aircraft artillery: "German Ago biplane of the Bayerische Flieger Abt. Nr. 9 Colmar. Very similar to and easily confused with French Voisin biplane. Color: whitish yellow. Transparent in the air. (Pay) Attention to iron crosses and oblique black and white stripes around the fuselage. (Bayer. Hauptstaatsarchiv Munich, Dept. 4, 6. Landwehr Division, Vol. 35, Folder 1).

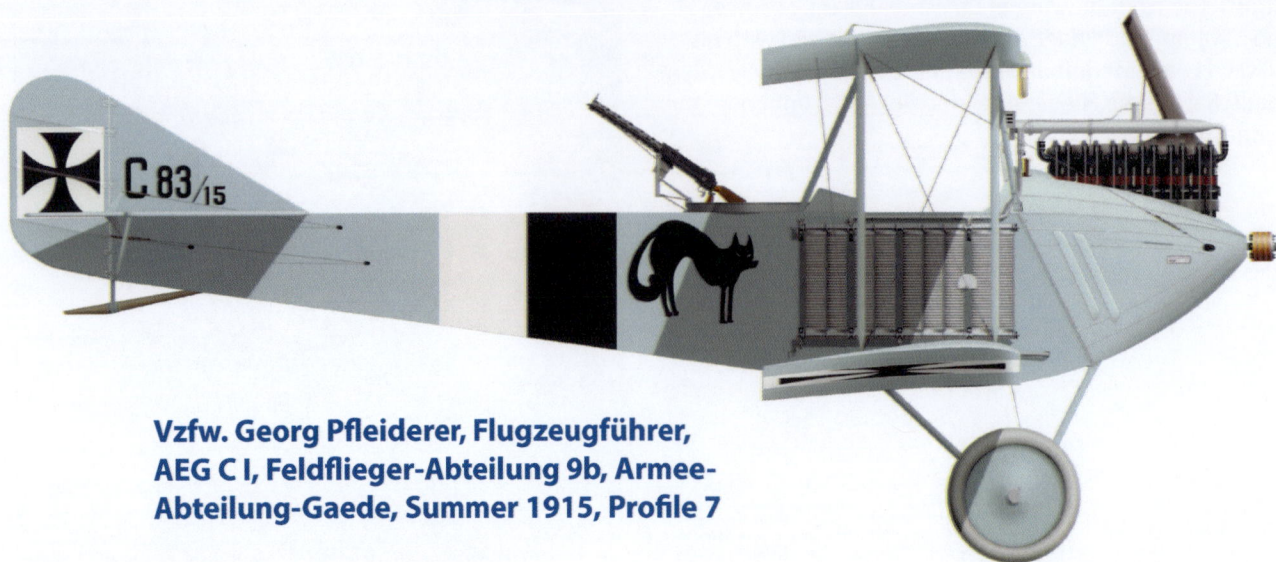

Vzfw. Georg Pfleiderer, Flugzeugführer, AEG C I, Feldflieger-Abteilung 9b, Armee-Abteilung-Gaede, Summer 1915, Profile 7

Heilbronn in the Kingdom of Württemberg. At the beginning of the war, he was a member of Field Artillery Regiment 29. He enlisted in the Fliegertruppe and was trained as an aircraft pilot with the Bavarian Flieger-Ersatz-Abteilung 1 as early as December 1914. After completing his training, he served his first frontline missions with Feldflieger-Abteilung 67 and later with Feldflieger-Abteilung 9b. He was promoted to Leutnant on 1 September 1915. He returned home in early 1917 and was assigned as a flight instructor in various Bavarian flying schools.

Promoted to Oberleutnant in March 1918, this was followed in August 1918 by his transfer as an Offizier z. b. V. to Fliegerabteilung (A) 289b, where he saw out the end of the war.[11]

Fokker Single-Seaters

Looking at the available photographs of Fokker single-seat fighters, it is apparent that the use of personal identifiers on these aircraft was rare.

Above: Blow-up of the cat on the fuselage of the AEG C I by Offz.Stellv. Georg Pfleiderer. Georg Pfleiderer is seen at left. (R. Kastner)

Brieftauben-Abteilung-Ostende

The available photos about the Brieftauben-Abteilung Ostende are unfortunately very sparse. However, as already mentioned, **there are several** letters **written by** Otto Parschau **to Anthony Fokker** during his **frontline** service, which give an interesting insight into the development of the early Fokker single seaters.[12]

Lt. Otto Parschau Fokker A III/M5K 16/15 (E I 3/15), Brieftauben-Abteilung Ostende, June 1915, Profile 8

The Fokker AIII/M5K 16/15 had a covering **with yellowish-sand colored** fabric. As a personal insignia, his Fokker had the lettering "Parschau" in black on the fuselage. In addition, the aircraft had a band with the Prussian colors black and white around the fuselage. This was to prevent confusion of his new

Lt. Otto Parschau Fokker A III/M5K 16/15 (E I 3/15), Brieftauben-Abteilung Ostende, June 1915, Profile 8

Left: Fokker A III/M5K 16/15 (E I 3/15) of Lt. Otto Parschau, Brieftauben-Abteilung-Ostende with the name "Lt. Parschau" and the Prussian colors black and white painted on the fuselage.

aircraft type by German aircrews with Allied aircraft such as the Morane-Saulnier M. This identification thus made his Fokker AIII/M5K 16/15 the first German "fighter" to carry **a personal insignia**!

Otto Paschau had a considerable influence on the development and use of the Fokker single-seater fighter. Accordingly, he was the first German airman to fly a Fokker single-seater fighter in combat in June 1915. Although he was never a member of a **Jagdstaffel, he** nevertheless had an important influence on the early German fighter arm.

Fokker E I, Feldflieger-Abteilung 42, summer 1915, profiles 9 and 9a (rudder)

Feldflieger-Abteilung 42 was formed in accordance with orders dated 13 September 1914 and was mobilized on 5 December 1914.[13] In the summer of 1915, the Abteilung was located at Bathelémont airfield, about 20 km east of Nancy, in the area of the Armee-Abteilung-Falkenhausen, on the French front.

A Fokker E I assigned to the department had a black rudder onto which a cigarette smoking skull with crossed bones was painted, and above it the inscription:

"Die Feinde werden nun den Fokker kennenlernen" (*"The enemies will now get to know the Fokker"*). On the side of the skull was

Above: The flag of the Kingdom of Prussia was often used by Prussian pilots as personal marking of their airplanes.

the saying: *"Mit viel Glück"* (*"with good luck"*), implying that they should consider themselves lucky to survive the encounter.

The aircraft was also marked with a slanted white and black fuselage band, with black and white trim, around the fuselage, as had been prescribed by the Armee-Abteilung Gaede since the spring of 1915. The reason for this may have been that from December 1914 to August 1915 the Armee-Abteilung Falkenhausen was subordinated to Armee-Abteilung-Gaede in tactical and operational matters, only then the Armee-Abteilung Falkenhausen became an independent Army. **Besides, this marking helped to avoid confusion with similar-looking French single-seaters.**

Above: Fokker E I of the Feldflieger-Abteilung 42 in the summer of 1915 in the area of Armee-Abteilung Falkenhausen. The aircraft carried a black and white fuselage band like the aircraft of Armee-Abteilung Gaede to avoid confusion with French aircraft. (G. VanWyngarden)

Bayerische Feldflieger-Abteilung 8

The Bavarian Feldflieger-Abteilung 8 was formed on 21 January 1915, at the Bavarian Flieger-Ersatz-Abteilung 1 in Oberschleißheim and was mobilized the same day.[14] In the autumn of 1915, the Abteilung was equipped with several Fokker and Pfalz single-seat fighters. The unit operated from airfield at Colmar in the area of the Armee-Abteilung Gaede.

The identification and painting of the aircraft are well documented by the large photo albums of former members Clemens Vollert and Erwin Wenig.

Lt. Erwin Wenig Fokker E II, Bavarian Feldflieger-Abteilung 8, November 1915, profile 10

After the black and white identification of the Army Department Gaede was no longer allowed to be used in the autumn of 1915, the airmen changed this identification. In some cases, the black-and-white band was additionally given another band in red, thus representing the German national colors black-white-red. Some Bavarian airmen, on the other hand, painted over the black band in blue, representing the Bavarian colors of white and blue. This did

not violate the order and resulted in simple way of providing the aircraft with a personal identification.

Erwin Wenig painted over the white part of the fuselage band in black. The black band thus represents his parent unit of the 1. Bayerischen Telegraphen-Bataillons (1st Bavarian Telegraph Battalion), which had black cap bands, black collars and black cuffs.[15] This is an early example of the regular unit's cap bands being used as a personal identifier for single-seat fighters.

A short biography of Erwin Wenig can be found in "*Jasta Colors Volume 1*".

2.2 Summary

Throughout 1915, the number of aircraft carrying individual marking gradually increased, although the majority of aircraft continued to carry no individual markings or identification. It is noticeable that quite a number of aircraft carried the Prussian colors black and white, usually applied as bands around the fuselage, and sometimes on the wheel covers. This may well have been an expression of patriotic sentiment on the part of native Prussian crews, but the primary reason was to be recognized by their

Fokker E I, Feldflieger-Abteilung 42, summer 1915, profiles 9 and 9a (rudder)

Left: "Blow up" of the rudder with the cigarette smoking skull and the inscriptions: "The enemies will now get to know the Fokker" and "with good luck." (G. VanWyngarden)

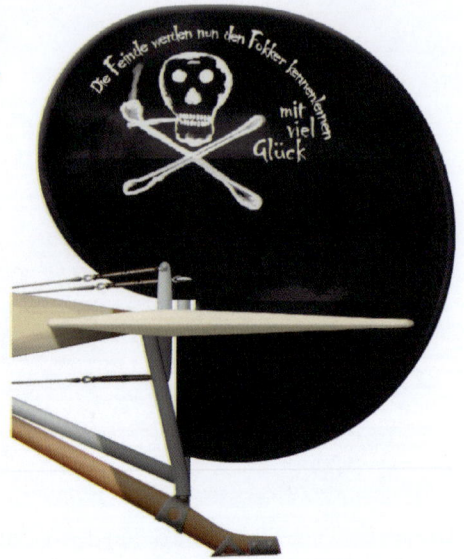

Below: Lt. Erwin Wenig in front of his Fokker E II of "Kampfeinsitzer Halbabteilung" (Single Seater-fighter half unit) of Bavarian Feldflieger-Abteilung 9 in November 1915. On the fuselage of the aircraft, the formerly black and white fuselage band is now uniformly black and represents the cap band of the pilot´s parent unit, the 1st Bavarian Telegraph Battalion.

Lt. Erwin Wenig Fokker E II, Bavarian Feldflieger-Abteilung 8, November 1915, profile 10

own troops, and an attempt to avoid confusion with Allied aircraft marked red, white and blue. As before, the aircrews feared that they could be mistaken for Allied aircraft and fired upon accordingly. For the same reason, black, white and red ribbons appeared on some of the German planes. Even though the crews of the anti-aircraft and ground MG crews had learned to distinguish between the silhouettes of Allied and German aircraft, German aircraft still found themselves subjected to anti-aircraft fire from their own ground defenses.

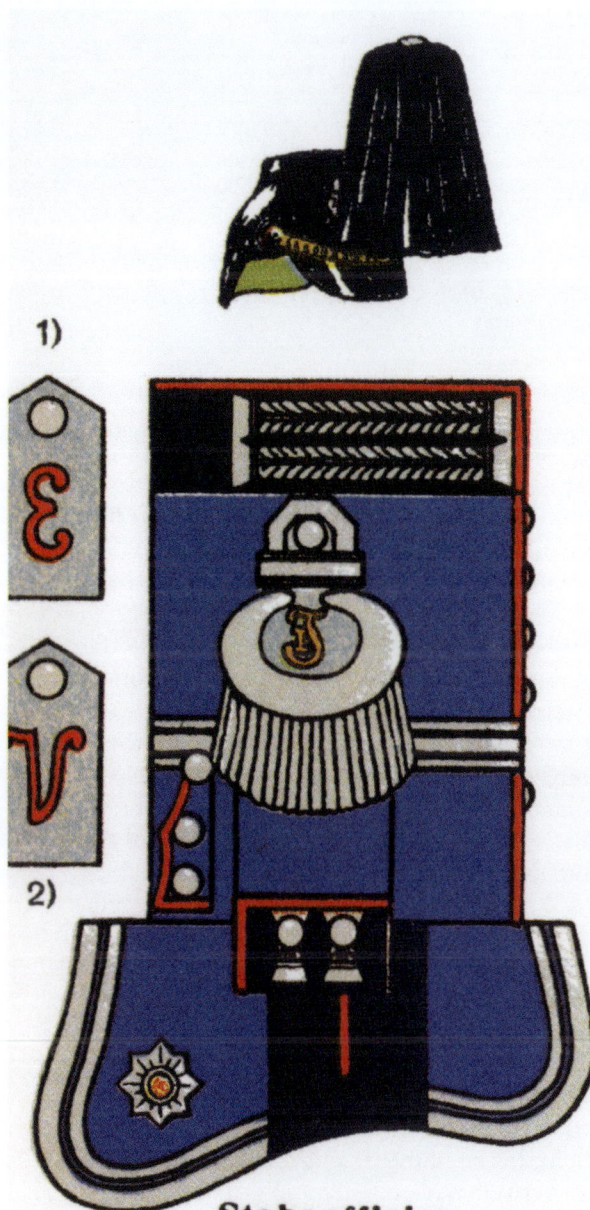

1)

2)

Stabsoffizier Parade Telegraphen-Batl. Nr. 1

Right: The peacetime uniform of the 1st Bavarian Telegraph Battalion. All telegraph battalions had black collars and black cuffs.

3. The German Fliegertruppe in 1916

Above: Five Fokker fighter seaters of the Kampfeinsitzer-Abteilung der Obersten Heeresleitung (Fokker single-seat-fighter-unit to protect the Supreme Army Command) in July 1916 stand ready for action on the airfield. In the foreground, Lt. Erwin Wenig.

On 1 April 1916, the German Fliegertruppe at the front had the following status: 17 Stabsoffiziere der Flieger, 81 Feldflieger-Abteilungen, and 27 Artillerieflieger-Abteilungen.[1] The "Brieftauben-Abteilung Ostende" (BAO), established on 27 November 1914, and the "Brieftauben-Abteilung-Metz" (BAM), established on 17 August 1915, had been converted on 20 December 1915 into the "Kampfgeschwader der Obersten Heeresleitung" (Fighting Squadrons of the Army High Command) I and II, each equipped with six Kampfstaffeln. By June 1916, another five Kampfgeschwader had been established, so that the Army High Command had now a total of seven Kampfgeschwader at their disposal.[2] Furthermore, the German Fliegertruppe had 17 Armee-Flug-Parks, 2 Riesenflugzeugabteilungen (Units with giant aeroplanes of the "R" category) Nos. 500 and 501, 2 experimental and training parks (Tergnier and Warsaw), Fliegerabteilung 300 "Pascha" in strength of 2 Feldflieger-Abteilungen in Palestine, and Fliegerverbände (aviation units) at the military mission in Turkey consisting of 3 Feldflieger-Abteilungen.[3]

Since the second half of 1915, German single-seater fighters had considerably limited the operational capabilities of Allied aviation units on the Western Front and inflicted corresponding losses on them. During the Battle of Verdun from February 1916 to June 1916, the period of superiority of the German Fokker single-seaters fighters, the "Fokker scourge" came to an end. They were opposed by new French fighters of the Nieuport XI type, grouped in "Escadrille de Chasse" (fighter squadrons). The German Fokker single- seaters, often still flying alone, had little prospects of success when facing this new opponent.

When the Battle of the Somme began on 1 July 1916, the German Fokker and Pfalz monoplane fighters, grouped in loose Kampfeinsitzer-Kommandos, were hopelessly outmmatched by the British "fighter squadrons" equipped with the new Airco D.H. 2. Allied aviators dominated the airspace above the battlefield. Allied artillery flying units were able to direct their artillery on German positions almost unchallenged, Allied planes strafed German trenches in low-level flight, and Allied bomber units carried their bomb loads deep into German-held areas. If German planes took off from their airfields to face the British units, they were driven back with losses. The artillerymen's ability to direct German gunfire at enemy positions was severely limited and bombing raids on Allied positions were almost impossible.

Facing this situation, the first German Jagdstaffeln were established at the front and successively equipped with the new Halberstadt, Fokker and Albatros fighter biplanes.

On 8 October 1916 came the decisive further step for the German air force. The entire air force and air

Above: The Nieuport XI of a French "Escadrille de Chasse". This type was far superior to the German E III and E IV and, together with the Airco DH 2, initiated the end of the "Fokker scourge".

defence resources of the Army, in the field and at home, were united under the "Kommandierenden General der Luftstreitkräfte" (Commanding General of the Air Forces). On 29 November 1916, the position of the "Stabsoffizier der Flieger" was changed to that of a "Kommandeur der Flieger" (Chief of all flying units assigned to an Army), who thus also had the corresponding command authority. (4) The position of the latter was presented in the "Weisungen für den Einsatz und die Verwendung von Fliegerverbände innerhalb einer Armee" (Instructions for the "Deployment and Use of Aviation Units Within an Army") published in May 1917, in the following words:

The Kommandeur der Flieger is the superior of all aviation units of an army; he makes proposals to the army high command for their distribution and employment.

He shall regulate and direct responsibly, in accordance with the instructions of the Army Headquarters, the employment of aviation units in the struggle for air supremacy.[5]

Further, the above instruction states:

"He drafts and submits to the Chief of Staff the orders for the use of the aviation units for the struggle for air supremacy, for long-range reconnaissance, for the uniform photographic survey of enemy positions and terrain, and for the conduct of special offensive actions. "[6]

The creation of this position represented a significant advance in the coordinated deployment of flying units to the front. As before with the Stabsoffizier der Flieger, so with the Kommandeur der Flieger, the performance and successes of the units under their command depended especially on their personalities. A number of Kommandeure der Flieger, such as Hptm. Helmuth Wilberg of the 4th Army, Hptm. Max Sorg of the 6th Army, and Hptm. Wilhelm Haehnelt of the 5th Army, had excellent reputations among the Fliegertruppe. They were known for their close and constructive cooperation with the aviation units assigned to their army. They knew how to use the experience of the units under their command and to represent the situation and even concerns of the flying units with commitment at higher levels.

Josef Jacobs, Staffelführer of Jagdstaffel 7 in 1917 with the German 4th Army, reported on his cooperation with his Kommandeur der Flieger Hptm. Helmuth Wilberg at the German 4th Army:

"Wilberg was not so much a superior as a friend, coordinator, contact person and supporter. He always had an open ear for all difficulties and supported us to the best of his ability. The successes of the (German) Jagdstaffeln in Flanders are also due to his leadership."[7]

In a few exceptional cases, the selection of the Kommandeur der Flieger proved to be a mistake,

Above: With the introduction of the Airco DH 2, the British Royal Flying Corps had a single-seater at their disposal that had superior flight characteristics compared to Fokker monoplanes. (C. Owers)

such as the inglorious appearance of Hptm. Bufe in June and July 1917 with the 4th Army. He was eventually replaced by Hptm. Wilberg due to the personal efforts of Manfred von Richthofen, Staffelführer der Jagdstaffel 11, and to the relief of the aviation units under his command.

For armies in main combat areas, the position of "Gruppenführer der Flieger" (group leader of aviation) was also created in the course of time at the Generalkommandos (Headquarters) of the army corps. His instructions considering the deployment of the flying units were formulated in "Weisung für den Einsatz und die Verwendung von Fliegerverbänden innerhalb einer Armee"

"1. On battle fronts requiring the employment of a large number of aviation units, the "Kommandeur der Flieger" will appoint a "Gruppenführer der Flieger" for the area of a group.

Generalleutnant von Hoeppner
Kommandierender General der Luftstreitkräfte.

Right: The commanding general of the German Fliegertruppe, Generalleutnant der Kavallerie Ernst von Hoeppner, proved to be a stroke of luck for the German Fliegertruppe.

2. *The "Gruppenführer der Flieger" is the superior of all aviation units of a group.*
3. *He regulates and directs, in accordance with the instructions of the Group (Corps) headquarter, the employment of aviation units for aerial reconnaissance and the struggle for air supremacy."*[8]

As a result of these directives, and the ever-increasing aircraft production in the homeland, ever larger numbers of aircraft were assigned to the individual aviation units. These units were then temporarily combined into the aforementioned groups.

In the course of time, the accumulation of aircraft inevitably led to the necessity in the air of being able to distinguish the aircraft of the different units.

3.1 Examples of the Markings of Aircraft in 1916

Artillerieflieger-Abteilung 211

Artillerieflieger-Abteilung 211 was formed on 23 October 1915 and was mobilized on 9 December 1915.[1] The Abteilung is documented by the photo album of the mechanic Uffz. Eberhardt Reiss, who also served in the same function with the Kampfstaffel Metz and the Fokkerstaffel Falkenhausen.

Aviatik C I, Artillerieflieger-Abteilung 211, Profile 11

The Aviatik C I had its fuselage and wings covered by a whitish linen. Additionally, the aircraft had a slanted ribbon in the German national colors of black-white-red on the rear fuselage. This is a good example of the use of the national colors as identification, because there was still the danger of being mistaken for an enemy aircraft especially by one's own air defence.[2]

Walter Böning, former member of the Bavarian Feldflieger-Abteilung 6, explained to me:

"A danger at that time was not only the enemy planes, but also the German anti-aircraft, especially when the silhouette of the aircraft differed little from French planes. We had with us the saying: 'The flak (anti-aircraft defense) knows neither friend nor foe, but only worthwhile targets'." It was not until the course of 1916 that the training of anti-aircraft soldiers was such that one could be largely sure of no longer being fired upon by German flak."[3]

Bavarian Feldflieger-Abteilung 6

The order to establish the Abteilung was issued on 6 November 1914, and on 12 December 1914, the unit was mobilized.[4] In the summer of 1916, the unit was located at the Bühl airfield near Saarburg in the area of Army Detachment A.[5]

Aircraft of this Feldflieger-Abteilung for the year 1916 are very well documented by the photos in the personal information of Dr. Walter Böning as well as by the photo albums of Walter Böning and Clemens Vollert. Walter Böning belonged to the Abteilung from 25 May 1916 to 26 November 1916, Clemens Vollert from 31 August 1916 to 11 July 1917.

Below: An Aviatik C I of Artillerieflieger-Abteilung 211 flipped over during landing. Such a landing was also disparagingly called "Damenlandung" (Ladies landing). The aircraft has a slanted band with the German national colors black-white-red as identification.

Aviatik C I, Artillerieflieger-Abteilung 211, Profile 11

Left: The German imperial flag black-white-red.

L.V.G. C I, Feldflieger-Abteilung 6b, Summer 1916, Profile 12 and 12a

The fuselage and wings of their L.V.G. C I aircraft were covered with a beige or a yellowish-bluish-gray fabric.[6] In this case the covering of the fuselage and the wings of the L.V.G. C I were interpreted as clear-doped linen fabric of a yellowish-beige color.

The crew of the aircraft carried a "winged Bavarian lion" as their personal insignia on the fuselage. The lion is the heraldic symbol of Bavaria and here the lion had now been given wings by the Bavarian Fliegertruppe. The "Bavarian" lion was most likely applied in the Bavarian national color blue.[7]

Feldflieger-Abteilung 18

The Abteilung had been formed in the course of mobilization on 1 August 1914 and assigned to the German 2nd Army. At the time, the Abteilungsführer was Hptm. Ernst Freiherr von Gersdorff.[8] In the summer of 1916 the Abteilung was stationed at Tourmignies Airfield, north of Douai, in the German 6th Army area.

Individual aircraft of the Abteilung are documented for this period by the photo album of Carl Allmenröder and photos in the collections of Reinhard Kastner and Reinhard Zankl.

L.V.G. C I, Feldflieger-Abteilung 6b, Summer 1916, Profile 12 and 12a

Below: This L.V.G. C I of the Bavarian Feldflieger-Abteilung 6 was marked with the winged Bavarian lion as personal marking on the fuselage, summer 1916.

Below: Enlargement of the "winged Bavarian lion".

Oblt. Ludwig Linck and Lt. Carl Allmenröder, pilot, Lt. Hein Adalbert Lubarsch and Oblt. Hans-Helmut von Boddien, observers, Roland C II, Feldflieger-Abteilung 18, October 1916, Profiles 13 and 13a

One of the most strikingly painted aircraft was the Roland C II of Oblt. Ludwig Linck, of Feldflieger-Abteilung 18. The Roland C II had a light bluish fabric finish of the fuselage and the wings. The aircraft initially had black checks painted on the light blue underside of the wings as personal identification and the name "*Meerkatzi*" (Little Guenon) on the vertical stabilizer.

Later, the aircraft received black diagonal stripes on the bluish fuselage as additional marking. According to Carl Allmenröder's notes, this Roland C II was also flown by him as a pilot. His preferred observers were Lt. Lubarsch, and Oblt. von Boddien.[9]

The aircraft is thus a good example of the fact that the individual painting of an aircraft could change over the course of its service and a photo is always only a momentary snapshot. So, it is up to the modeler which version of the painting scheme of this aircraft he want to represent.

In 1974 I came in contact with the aviation historian Gero von Langsdorff, son of the pilot, aviation historian, and author Werner von Langsdorff who had written a number of books about German World War I aviation in the 1930s. Herr Gero von Langsdorff not only supported me with documents and photos from his father's archive but put me also in contact with Herrn Ludwig Zacharias from Cologne who was one of the aviation enthusiasts of the first generation. He had been in contact with former German fighter pilots as early as the 1930s. Herr Ludwig Zacharias was not only a very prolific historian but also a very friendly and helpful person who kindly provided me with the result of his research and also with addresses of former German fighter pilots he had been in contact with.

One of his long-time contacts was Wilhelm Allmenröder, brother of the Pour le Mérite recipient Carl Allmenröder. Herr Zacharias gave me the address of the widow of Wilhelm Allmenröder, Frau Helene Allmenröder, who lived in Prien at

Above: The Royal Bavarian Coat of Arms.

Above: A variation of the "Bavarian Lion" was and is still used by the Munich sports-club 1860 München. The winged lion on the plane is applied in a very similar style.

the Chiemsee Lake east of Munich. After the first telephone contact Frau Allmenröder invited me to visit her. As I told her that I would come with my girlfriend she had pastries and coffee prepared for us and it became a very nice meeting.

During my visit Frau Allmenröder told me that she had been the fiancée of Carl Allmenröder. After his death, his brother Wilhelm took care of her, eventually both fell in love and married.

Besides the personal memories and letters of Carl Allmenröder, Frau Allmenröder presented two photo albums of Carl and Wilhelm that I was able to photograph. Among the photos I came across the strikingly painted Roland C II of Feldflieger-Abteilung 18, which was later supplemented

by additional photos and information of this machine. She loaned me the photo albums of the Allmenröder brothers and many documents except for the personal letters she had gotten from Carl. Nevertheless, she was so kind to provide information of the military content of the letters. We had two additional very nice meetings with Frau Allmenröder. During one of the visits, she told of her great disappointment that after the Second World War a street in Carl Allmenröder's hometown Wald that had been named after him had been renamed by local politicians under the excuse that the Nazis had used him for their propaganda! The fact that Carl Allmenröder had already been dead for 16 years at the time of the Nazi seizure of power and that he could therefore never have had anything to do with the Nazis was completely unimportant for those politicians. The only important thing for them was to make a name for themselves and to buster their own political career!

Oblt. Ludwig Linck. Only little information is available about him. He was born in Bremen on 19 October 1889.[10] After serving with Feldflieger-Abteilung 18, he was appointed Staffelführer of the newly established Jagdstaffel Linck on 25 September 1916, which was etatised as Jagdstaffel 10 on 6 October 1916. A short time later, on 22 October 1916, he was killed in air combat.[11]

Meerkatzi

Oblt. Ludwig Linck and Lt. Carl Allmenröder, pilot, Lt. Hein Adalbert Lubarsch and Oblt. Hans-Helmut von Boddien, observers, Roland C II, Feldflieger-Abteilung 18, October 1916, Profiles 13 and 13a

Above: Initially, only the lower wings of the Roland C II of Oblt. Ludwig Linck were painted with a black checkerboard pattern, and the inscription "Meerkatzi" (type of mongoose) was applied onto the tail fin. Note that the forward fuselage side window has been opened. (R. Zankl)

Carl Allmenröder was born on 3 May 1986, in Wald near Solingen, the son of a pastor. A medical student at the outbreak of war, he went into the field as a volunteer with Field Artillery Regiment No. 62. At the end of 1914 he was briefly with Regiment No. 20 but returned to his old unit in January 1915. Mainly in action in Galicia, he was promoted to Leutnant on 30 March 1915, and awarded E.K. II (about April 1915) and the Friedrich-August-Kreuz 1st Class (August 1915). In March 1916 he transferred to the Fliegertruppe and received his training as a pilot from 29 March at FEA 5 in Hanover and the Flaying school Halberstadt.

After completing his training, he was commanded to Armee-Flug-Park 6 on 25 August and was transferred to Feldflieger-Abteilung 18 a month later.

Right: Lt. Carl Allmenröder as an artillery officer prior to his training as a pilot.

Above: I: Oblt. Ludwig Linck in front of his Roland C II, which has now received additional markings in the form of black diagonal stripes. A captured British "Lewis" machine gun has been fitted. (T. Weber)

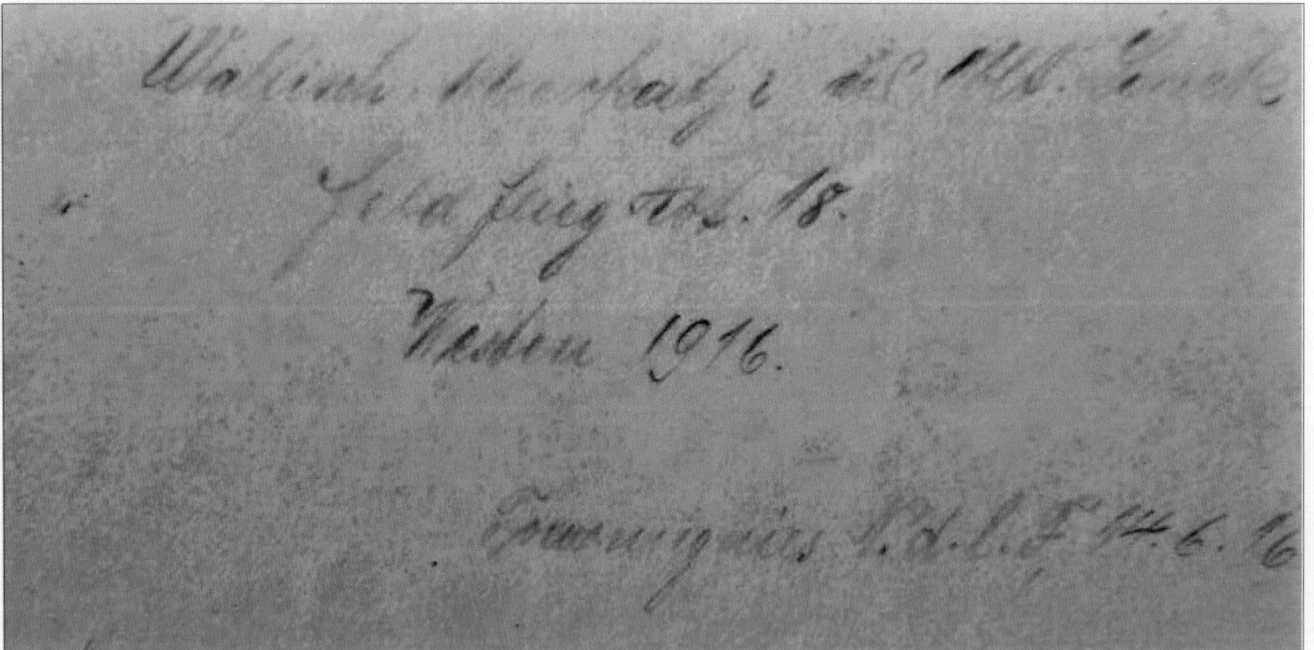

Above: The inscription on the backside of the photo (top) reads: "Walfisch Merkatzi of Oblt. Link, West 1916. Feldflieger-Abteilung 18, Tourmignies, N.d.l.F 16.6.16." (R. Zankl)

When the first Jagdstaffeln were formed in the area of the German 6th Army in the fall, he received his transfer order to Jagdstaffel 11 commanded by Oblt. Rudolf Lang on 20 November 1916. Like all other pilots of this Jagdstaffel he needed a relatively long start-up time until he could finally achieve his 1st aerial victory on 16 February 1917.

Under the leadership of Manfred von Richthofen, however, both the success of the Jagdstaffel and his successes went steeply uphill: On 24 March, he was awarded the E.K. I after scoring 4 aerial victories, and by the end of April he had already achieved 9 kills. His most successful month, however, was May 1917 with a total of 13 aerial victories. When he was awarded the Knight's Cross with Swords of the Royal House Order of Hohenzollern on 9 June, he already

Above: The Roland C II "Meerkatzi" was also flown by Lt. Carl Allmenröder and his observers Adalbert Lubarsch, and Oblt. Hans-Helmuth von Boddien, respectively. In front of the aircraft is Oblt. Hans-Helmuth von Boddien.

had 26 aerial victories, and only five days later he was awarded the Pour le Mérite. But just two weeks later, his meteoric fighter flying career came to an end. On 27 June 1917, his Albatros D V crashed, breaking up in midair after being hit by flak while flying over the lines, and it came down in no man's land near Klein-Zillebeke at 8:20. His body was recovered by a German assault team during the night of 28–29 June and was transferred home. On 20 July 1917, he was posthumously awarded the Bavarian

Military Order of Merit IV. Class with Crown.[12]

Hein Adalbert Lubartsch was born in Rostock on 4 November 1895. He went to war with the Kurmärkisches Dragoner Regiment No. 14. He enlisted in the Fliegertruppe and after his training joined the Feldflieger-Abteilung 18. Like his friend Carl Allmenröder, he did not survive the war. He was killed in air combat on 17 September 1917.[13]

Above: Lineup of Rumpler C I of Kampfstaffel 2 of Kampfgeschwader I running up their engines for take-off. The Roman numeral "II" indicates that the planes belonged to Kampfstaffel II, while the Arabic numerals 1 through 7 were the sequential identification numbers of the individual aircraft. With seven aircraft, the squadron exceeded the nominal strength of six aircraft by one. Note the different styles of Iron Cross national markings on the aircraft. (R. Kastner)

Above: Oblt. Hans Ulrich von Trotha in the observer's seat waves to another Kagohl I aircraft. The L.V.G. C I of Kampfstaffel IV with the crew Lt. Rolf von Lersner, pilot and Oblt. Hans Ulrich von Trotha, observer, was marked with the colors of the coat of arms of the von Lersner family or the von Trotta family. (T. Philips)

Above: The A.E.G. G III flown by Lt. Rolf von Lersner was marked with the same colored fuselage bands as personal identification as the L.V.G. C I which he flew together with Oblt. von Trotha. The photo caption reads: A.E.G with coat of arms color.

Hans-Helmut von Boddien. Almost no data are known about his early military career; except that he served with the Kürassier-Regiment "Königin" No. 2 in 1914/15 and passed his officer's examination on 18 October 1915. From 1916 he flew as an observer with Feldflieger-Abteilung 18. Trained as a pilot in early 1917, he was transferred from Jagdstaffelschule I in Valenciennes to Jagdstaffel 11, the "Richthofenstaffel," on 24 June. There he made relatively few flights against the enemy during the next six months, as he was heavily involved into taking care or administrative duties of the unit. On 26 February 1918, he was appointed Staffelführer of the newly formed Jagdstaffel 59 and was able to achieve his first aerial victory in the following month. After receiving Fokker D VIIs, he was able to add another four aerial victories in August and September. On 27 September, he was wounded in air combat with D.H. 9s by a shot in the calf and saw out the end of the war in hospital.

He then enlisted in the German Border Guard in the east and flew with Fliegerabteilung 424 during the fights in the Baltics. Here he did not return from a flight on 29 November 1919 and has since been considered missing in action.[14]

Kampfgeschwader der Obersten Heeresleitung

Unlike the Fliegerabteilungen and the pilots of the Fokker and Pfalz single-seaters, the Kampfgeschwader flew most of their missions in close formation. This meant that each aircrew had to be able to recognize the position of the other aircraft.

The coat of arms of the von Lersner family. The ancestral colors of the coat of arms are yellow and red.

The black and yellow coat of arms of the von Trotha family.

Above: Lt. Rolf Freiherr von Lersner, pilot in Kagohl I, Kampfstaffel IV.

Right: Oblt. Hans-Ulrich von Trotta in front of an L.V.G. C I of Kampfgeschwader I.

Kampfgeschwader der Obersten Heeresleitung I

The presentation of the identification of the aircraft of Kampfgeschwader I is based on the photo albums, notes and documents of the unit members Franz Walz, Waldemar Christiansen, and Rudolf von Lersner.[15] According to these sources, the aircraft bore the Latin numbers of Kampfstaffel I through VI, in addition to Arabic numerals 1 through 6 that were applied to the fuselage sides.[16]

The general interpretation is that the numbers were painted in black, and I have adopted this since I have no testimony which indicates the contrary.

The fuselages and the wings of the L.V.G. C II were covered with a light sand coloured linen on which the – presumably black – numbers would have been clearly visible.

Lt. Rolf von Lersner, Oblt. Hans Ulrich von Trotha L.V.G. C I , Kagohl I, Kampfstaffel 4, Profiles 14 (von Lersner colors), 14a (von Trotha colors)

The aircraft has the identification "IV" for Kampfstaffel 4 and the individual identification "1" for the aircraft. In addition, the aircraft carried a two-colored slanted band on the fuselage. This band was described in previous publications as black and yellow, the colors of the von Trotha family. However, a note in Rolf von Lersner's photo album suggests that the band was possibly red and yellow, which were the von Lersners' heraldic colors. Rolf von Lersner later had an identical band marked on the fuselage of his A.E.G. G I, and this band was verifiably red and yellow.[17]

Lt. Rolf von Lersner, Oblt. Hans Ulrich von Trotha L.V.G. C I , Kagohl I, Kampfstaffel 4, Profiles 14 (von Lersner colors), 14a (von Trotha colors)

In 1982 I came into contact with the von Lersner family. Herr Albert von Lersner wrote to me on 21 May 1982:

Dear Herr Schmäling,
In response to your letter of the 11th of this month concerning my cousin Rolf (not Rudolf) von Lersner, I can inform you of the following:
Rolf, Adolar, Adolf was born on 26 December 1893 in Deutz/Rhine and initially received his education at home before entering the cadet school in Potsdam and Großlichterfelde. In the spring of 1912, he was commissioned as an ensign in the Guard Cuirassier Regiment in Berlin. After attending war school in Danzig, he was recruited into the Pomeranian Dragoon Regiment No. 11 in Lyck in 1913. He did not stay there for long. His venturesome spirit led him to abandon his military career and travel to Argentina. Before the outbreak of the World War, he managed to return to Germany in time, where he was initially accepted as an ensign in the Dragoon Regiment No. 11. After a short induction, he was soon sent to the front in Russia, soon became an officer and, after healing from a minor wound, trained as an aviation officer. He soon gained a reputation as an able aviator. He made over 1,000 enemy flights in Russia, Romania and on the Western Front.[18]

Rolf von Lersner joined Kampfgeschwader I after completing his training. On 8 April 1917, he was transferred to Jagdstaffel Boelcke. On 9 May 1917, he, together with Lt. Werner Voss, shot down a Sopwith Pup near Lesdain. The aerial victory was awarded to Lt. Voss. On 11 May 1917, he reported shooting down a B.E. biplane, but again he would not receive confirmation for his claim. The date of his transfer from Jagdstaffel Boelcke is not known; his name does not appear in the list of pilots dated 1 August 1917.[19] According to the family, he crashed fatally during a test flight on 25 August 1917. He was married to Margarete von Zitzewitz and left no descendants.[20]

Hans-Ulrich von Trotha was born on 1 July 1889, in Potsdam, a member of an old Saxon noble family. He met a fate similar to that of his aircraft pilot Rolf von Lersner. He accompanied his Geschwaderführer Hptm. Ernst Brandenburg, to the Supreme Headquarters at Bad Kreuznach, where Brandenburg was awarded the Order Pour le Mérite. On takeoff for the return flight on 19 June 1917, the aircraft crashed. Hptm. Brandenburg survived, suffering serious injuries. The unlucky Oblt. von Trotha, however, was killed in the crash.

Above: A contemporary postcard of Kampfstaffel 8 in Kampfgeschwader II showing the individual identifications applied to the seven D.F.W C V of the unit. The aircraft of the Staffelführer is presented as the main image, appearing in the shape of a black airplane silhouette. A Prussian cockade in black and white has been placed above the "OHL" inscription. (T. Weber)

He found his final resting place near the family castle. At that time Hans-Ulrich von Trotha was leader of Bombenstaffel (Bombing echelon) 15 in the "Bombengeschwader (Bombing Squadron) III der Obersten Heeresleitung"[21].

Kampfgeschwader der Obersten Heeresleitung II, Kampfstaffel 8 und 10

As already mentioned in the preface, I have repeatedly had the experience that a single photograph brings new aspects to the research of the identification and painting of German air units. This

Above: Lt. Erwin Böhme and his observer Lt. Lademacher in front of their Albatros C III 766/16. The personal marking applied to the right side of the fuselage was a large dragon. The wooden fuselage of the aircraft appears dark on the photos due to the lighting conditions, which lead to the wrong interpretation in some publication that the fuselage of this particular aircraft was rust red.

Below: The crew Lt. Böhme and Lt. Lademacher after a successful front-line flight. The personal marking on the left side of the fuselage was a crocodile. Photographed under different lighting conditions, the fuselage appears in a much lighter shade and proves that the fuselage was clear varnished wood.

Above: The aircraft of Kampfstaffel 10 of Kampfgeschwader II also had individual symbols on the fuselages, such as a dragon, a crescent and colored disk.

Albatros C III Kagohl II Kampfstaffel 10, Profile 15

was the case when my friend and colleague Tobias Weber was able to acquire a postcard of Kampfstaffel 8 in Kampfgeschwader II. On the card, the insignia of the six aircraft of Kampfstaffel 8 are depicted as pennants. Additionally, a D.F.W. C V is shown on the postcard.

What is special about this illustration is that it is a very early documentation where all aircraft of a German aviation unit were marked with individual symbols. Also, the other Kampfstaffeln of Kagohl II had – based on the available documents – different symbols as identifiers. With these symbols the identification of the airplanes of Kagohl II differs

from the number identifiers used for example by Kagohls I, IV and VI.

It is interesting in this context that similar symbols were later used by the pilots of the Jagdstaffeln. An example of this is Jagdstaffel 37 under the command of Staffelführer Oblt. Kurt Grasshoff in the summer of 1917, which will be the subject of a later volume of the Jasta Color series.

Unfortunately, the authors don't know of any photo of a D.F.W. C V of Kampfstaffel 8. However, in the photo album of Erwin Böhme there were a number of photos of Kampfstaffel 10. Since Kampfstaffel 10 also used symbols for the individual identification of their aircraft, an Albatros C III of this unit is presented as an example of such an identification used by Kampfgeschwader II.

Albatros C III Kagohl II Kampfstaffel 10, Profile 15

The aircraft had as its individual identification of a black crescent moon outlined in white on the wooden fuselage glazed with clear varnish and a black crescent moon on the white-painted fin. The hubcaps are painted half black and white. Since this style of painting of the hubcaps is photographically documented for a number of aircraft of Kampfstaffel 10 it may have been the Staffel identification.

According to the research of Jörn Leckscheid, all Albatros C.III of the 7xx/15 lot had "light" varnished fuselages made of birch plywood panels that were locked against each other in three layers and coated in a glossy varnish at the factory. In the photo, the wooden fuselage of this plane appears dark because of its angle to the camera, but it actually is light yellowish wood.

How the angle of the camera and light conditions can alter the grey scale on orthochromatic films can be seen on the different photos of the famous Albatros C III with the "Dragon and Crocodile" marking, flown by Erwin Böhme. In one of the photos the fuselage appears dark grey which led, in recent publications, to the conclusion that the fuselage was rust-red. But the other photos prove that the fuselage was covered with a light natural yellowish plywood. This serves as an additional example that relying solely on the so called "grey-scale interpretation" simply does not work!

Right: The original document of Kampfgeschwader III shows the marking system of Kampfstaffel 13 and 14. (R. Kastner).

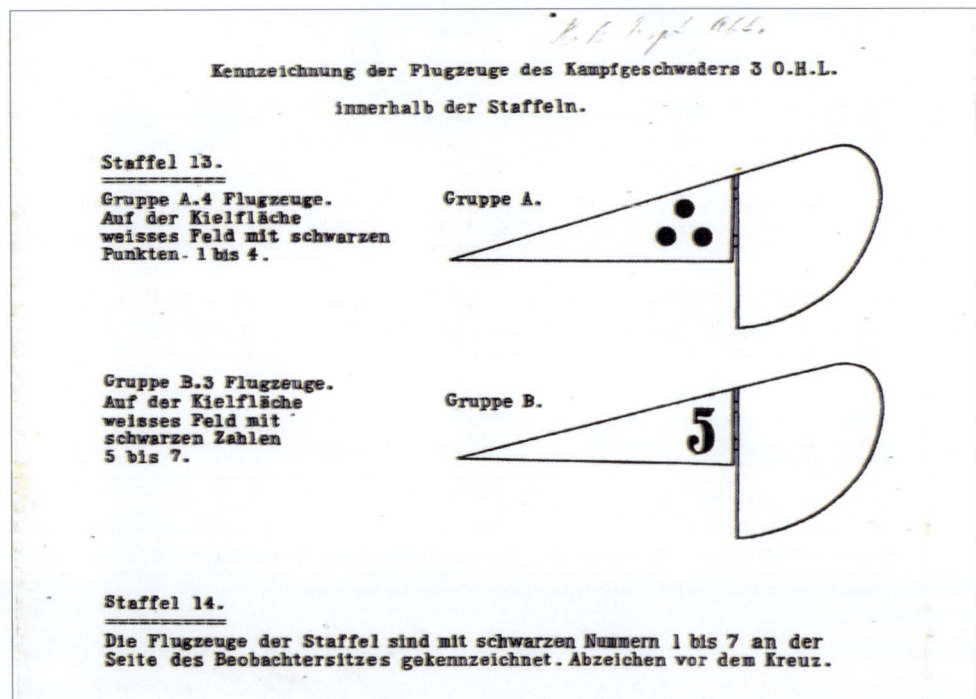

Right: The markings of Kampfstaffel 15 of Kampfgeschwader III. (R. Kastner).

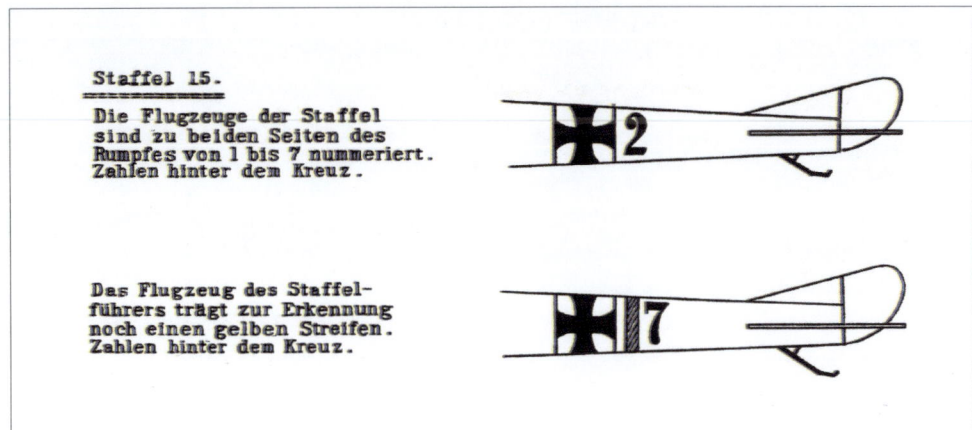

Staffel 16.

Erkennungszeichen der
Staffel: schwarzer Streifen
auf weissem Grunde.
Zeichen hinter dem Kreuz.

Erkennungszeichen des
Flugzeuges vom Staffel-
führer : schwarzer Streifen
in T-Form auf weissem
Grunde.
Zeichen hinter dem Kreuz.

Left: The markings of Kampfstaffel 16 of Kampfgeschwader III. (R. Kastner).

Staffel 17.

Erkennungszeichen der Flugzeuge
der Staffel: zwei gekreuzte schwarze
Streifen hinter dem Kreuz.
Das Zeichen auf allen Seiten
des Rumpfes u. der Räder.

Erkennungszeichen des Flugzeuges vom Staffelführer:
über dem gekreuzten Streifen
ein wagerechter.
Das Zeichen auf allen Seiten
des Rumpfes u. der Räder.

Staffel 18.

V=ähnliches Zeichen auf allen
Seiten des Rumpfes und Aussenseiten der Räder. Abzeichen
vor dem Kreuz.

Erkennungszeichen des Flugzeuges
vom Staffelführer: über dem V=
ähnlichen Zeichen ein wagerechter
Streifen.
Abzeichen auf allen Seiten des
Rumpfes und Aussenseiten der
Räder.
Für die weitere Kennzeichnung des Führerflugzeuges ist ein kleines
Spiegelprisma in Vorbereitung, welches – auf der oberen Tragfläche ange-
bracht – sich im Fluge dreht und Lichtreflexe verursacht.

Verb.Off.K.G.3. O.H.L.
empf. 2/5.16. Nr.86

                An
                        A.O.K.6

        Mit der Bitte, die Kennzeichen der Armee zur Kenntnis zu

bringen.

                        A.H.Qu., den 2.5.16
                        I.A.d.K.G.O.H.L.

Left: The markings of Kampfstaffel 17 and 18 of the Kampfgeschwader III. (R. Kastner).

Above: The L.V.G. C II of Kampfstaffel 16 of the crew Vzfw. Arthur Schorisch and Lt. Karl Meierdirks. Both later retrained as fighter pilots and were assigned to Jagdstaffel 12. (T. Philips)
Below: The L.V.G. C II of Kampfstaffel 16 of the crew Vzfw. Arthur Schorisch and Lt. Karl Meierdirks. Both later retrained as fighter pilots and were assigned to Jagdstaffel 12. (T. Philips)

Above: The L.V.G. C II of Kampfstaffel 16 of the crew Vzfw. Arthur Schorisch and Lt. Karl Meierdirks. Both later retrained as fighter pilots and were assigned to Jagdstaffel 12. (T. Philips)

Above: The L.V.G. C II of the crew Lt. Heinrich Geigl, pilot and Lt. Alwin Kinkelin, observer, Kampfstaffel 36, July/August 1916. The aircraft is marked with a black or red zig-zag fuselage band as identification of Kampfstaffel 36 and the number "5" as individual identification of the aircraft.

Kampfgeschwader der Obersten Heeresleitung III

Written instructions for the identification of a German aviation unit or the individual identification of aircraft of a Staffel exist only in some very rare cases. My friend and colleague Reinhard Kastner is in possession of one of these very rare documents. It shows that the Kampfgeschwader III had a fixed identification for each of the 6 Kampfstaffeln and these in turn were divided into half Staffeln of 3 aircraft each. The division into half-Staffeln may have served as a model for the later "Ketten" of three aircraft of the Jagdstaffeln.

Kampfstaffel 13

Group A: On the fin white field with black dots (Aircraft 1: one dot, Aircraft 2: two dots, Aircraft 3: three dots, Aircraft 4: four dots)
Group B: On the fin white field with black numbers 5 to 7. Kampfstaffel 13 was equipped with seven instead of the six budgetary aircraft.

Kampfstaffel 14

The aircraft are marked with black numbers 1 through 7 on the side of the observer's seat. Numbers in front of the cross.

Kampfstaffel 15

The aircraft are numbered from 1 to 7 on both sides of the fuselage, number behind the cross.

The aircraft of the Staffelführer additionally bore a yellow stripe for identification. Number behind the cross.

Kampfstaffel 16

Marking: Black stripe on white background. Marking behind the cross.

Identification mark of the aircraft of the Staffelführer: black stripe in T-shape on white background. Marking behind the cross.

Kampfstaffel 17

Marking of the aircraft: Two crossed black stripes behind the cross. The marking on all sides of the fuselage and wheels.

Identification of the aircraft of the Staffelführer: Above the crossed stripes a horizontal one. The marking on all sides of the fuselage and wheels

Kampfstaffel 18

Marking: V-like mark on all sides of the fuselage and outer sides of the wheels. Marking in front of the cross.

Identification of the aircraft of the Staffelführer: above V-like sign a horizontal stripe. The sign on all sides of the fuselage and outside of the wheels.

For further marking of the aircraft of the Staffelführer a small mirror prism was in preparation which rotated in flight and caused light reflections.

Kampfeinsitzer- und Fokker-Staffeln

Even in 1916, the majority of Fokker and Pfalz monoplane fighters flew without personal markings. However, some of these fighter planes still had black and white ribbons or bands as markings to avoid confusion with similar looking Allied aircraft.

Above: Lt. Josef Jacobs and his mechanics pose in front of his green-brown mottled Fokker E III (probably E 339/16), Fokkerstaffel West, June 1916.

Josef Jacobs, Fokker E III (probably E 339/16), Fokkerstaffel West, June 1916, Profile 16

Josef Jacobs, Fokker E III (probably E 339/16), Fokkerstaffel West, June 1916, Profile 16

When delivered from the factory, Josef Jacobs´ Fokker E III was covered with a beige linen fabric. His mechanics added a camouflage scheme that was overall predominantly green, but also had brown-colored spots mixed in. As Josef Jacobs told me, this had the advantage that he could not be easily recognized in flight by aircraft in a higher position, as the green camouflage with brown spots was not so easy to distinguish from the landscape below. Another advantage was that the aircraft could be well camouflaged on the ground, providing some protection against bombing by enemy aircraft. He explained that his ground crew camouflaged the Fokker with twigs and hay in the summer to avoid detection from the air by enemy planes. This was one of the early "front line camouflage schemes" of a single-seat fighter aircraft.[22]

A short biography of Josef Jacobs appeared in *Jasta Colors* Volume 1, pages 61–62.

Lt. Walter von Bülow, Pfalz E II, Fliegerabteilung 300 „Pascha", Palestine Spring/Summer 1916, Profile 17

His Pfalz E II had as a personal insignia a skull with two crossed bones on the engine hood.[23] This was the emblem of his parent unit, the famous Brunswick Hussar Regiment 17. The tradition of this Hussar regiment went back to the time of the war against Napoleon, and it was considered an honour to be a member of this regiment. The fuselage and the wings of the Pfalz had the usual whitish fabric covering with the black edgings. This is another early example of how the emblem of the parent regiment of the pilot was transferred to his aircraft as his personal insignia.

In 1984, after a short correspondence, I was invited by Freifrau Rosa von Bülow, the widow of Harry von Bülow, Walter's brother, to visit her in Alt-Bockhorst in the Emsland in north-west Germany. Soon I was on the way with my girlfriend Kerstin. What was planned as a short visit became a full day stay with a delightful hostess, ending late

Below: Fokker E IV of Kampfeinsitzer-Kommando III in the area of the German 6th Army has black and white crossed bands as personal identification. The wheel covers have also been painted in black and white.

Above: Lt. Walter von Bülow standing on the right in front of his Pfalz E II at Fliegerabteilung 300 "Pascha" in Palestine, to the left his mechanics. An unknown airman the takes up the pilot's seat. The emblem of the Brunswick Hussar Regiment 17 has been marked on the engine cowling.

Above: Magnification of the marking applied to the engine cover of Lt. Walter von Bülow's Pfalz single-seater.

Above: Lt. Walter von Bülow in front of the hangar tent that housed his Pfalz E II, Fliegerabteilung 300 "Pascha".

in the evening. Mrs. von Bülow had prepared all the photos and documents so that I could copy them. She also showed us the 1/72 scale airplane models made by a grandson in the schemes of the three von Bülow brothers, Walter, Harry and Konrad von Bülow. As a parting gift, Mrs. von Bülow handed me two postcards with the signatures of the participants of the "Bothkamper Fliegertage" (Bothkamper pilot's days) from the years 1924 and 1925. As she told me, her late husband Harry von Bülow organized

meetings of former aviation comrades of the First World War in Bothkamp every year in the summer. A good number of the World War I pilots were guests, and some of them, such as Paul Bäumer, came in their own aircraft.

Walter von Bülow-Bothkamp was born in Eckernförde in Holstein on 25 April 1894. He was a law student at the famous University of Heidelberg at the outbreak of the war and immediately enlisted

Lt. Walter von Bülow, Pfalz E II, Fliegerabteilung 300 „Pascha", Palestine spring/Summer 1916, Profile 17

Above: Lt. Walter von Bülow in the cockpit of a Roland C II at the "Versuchspark Tergnier".

Leutnant Dienstanzug

Above: The peacetime uniform of Brunswick Hussar Regiment No. 17.

Above: The fur cap of Brunswick Hussar Regiment No. 17 with the skull emblem. This insignia was used by Lt. von Bülow on the Pfalz E II he flew while serving with Fliegerabteilung 300 "Pascha". (Brunswick State Museum)

in the army as a volunteer. During the first months of the war, he fought with the Brunswick Hussar Regiment No. 17 in Alsace and was also promoted to Leutnant there. In the spring of 1915, he enlisted in the Fliegertruppe and, after training as a pilot with Flieger-Ersatz-Abteilung 5 in Hanover, joined Feldflieger-Abteilung 22 in Champagne in September. There he was assigned one of the very first Fokker monoplanes. As early as October 1915

he was able to achieve two aerial victories and was awarded the E.K. I for this.

On 14 March 1916, he was transferred to Fliegerabteilung 300 "Pascha" in Palestine, where he was able to achieve two more air victories over the Sinai in the summer flying a two-seater Rumpler C I. Furthermore, during the "Sea Battle off El Arish", he successfully fought a British warship unit with bombs and on-board weapons. In addition to the Rumpler C I, he flew a good number of missions with his "skull-marked" Pfalz E II.

In November 1916 he returned to Germany and flew since 7 December with the newly formed Jagdstaffel 18 in Flanders. Here he was able to

Above: Invitation card of the "Pilotenfest Bothkamp 11. July 1925.

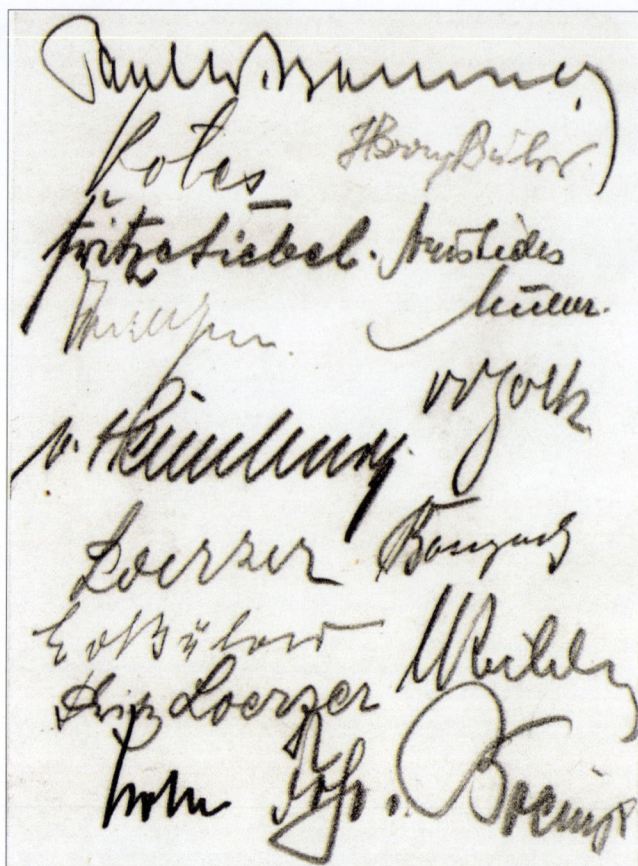

Above: Signed backside with the participants. The following signs are identified: Paul Bäumer, Josef Jacobs (Köbes) Harry von Bülow, Fritz Liebel, Aristide Müller, von York, von Heimburg, Bruno Loerzer, Heinz Bongartz, E. von Bülow, Fritz Loerzer, Frhr. von Boenigk

increase his number of victories to 13 in a short time and was awarded the Knight's Cross of the Royal House Order of Hohenzollern on 24 April 1917. On 15 May, he was appointed Staffelführer of Jagdstaffel 36. In June, the Staffel moved from Champagne to Flanders, where he led the Jagdstaffel to great successes and was able to seamlessly continue his previous successes.

By the end of September, he already had 21 aerial victories, and three weeks later, on 20 October 1917, he was awarded the Order Pour le Mérite after his 22nd victory. By December, he had increased his number of aerial victories to 28 and was then appointed Staffelführer of Jagdstaffel "Boelcke" on 13 December, succeeding Erwin Boehme. But only three weeks later, on 6 January 1918, his career ended. In aerial combat with SPADs and Sopwiths, he pursued an enemy alone across the lines, was caught in the rear by other opponents, and finally crashed fatally on Albatros D V 2080/17 between St. Julien and Paschendaele.[24]

Kampfeinsitzer-Kommando Ensisheim
Based on available photographs no Staffel marking or identification system of the different aircraft of the Kampfeinsitzer-Kommando or Fokkerstaffeln is verifiable, with one exception![25]

This single known exception was the Kampfeinsitzer-Kommando Ensisheim in the summer of 1916.

Hans Sippel, a mechanic of the Kampfeinsitzer-Kommando and later of Jagdstaffel 16, was in correspondence with Dr. Gustav Bock for quite some time. In one of his letters to Dr. Bock, he wrote:

"The fuselages of our "fighter planes" of KEK Ensisheim were painted light grasshopper green, which gave us the nickname "Grashüpfer Staffel" (Grasshopper unit). When we appeared somewhere it was only said: "The Grasshoppers are coming".[26]

KEK Ensisheim was at this time equipped with Fokker D II and some old Fokker E III and E IV aircraft. The Fokker D IIs that were delivered to the Kampfeinsitzer-Kommando originally carried an all-over light-colored linen scheme, although some already carried the new two-color camouflage

Above: Four Fokker D IIs and single examples of the Fokker E III and E IV of Kampfeinsitzer-Kommando-Ensisheim, commonly called the "Grasshopper Staffel" by nearby units, are lined up in parade formation for the benefit of the photographer. (W. Bock)

scheme on the upper surfaces, which had just been introduced at the Fokker factory. Soon after arriving at the unit, a number of aircraft were painted in light grasshopper green. Additionally grim looking faces were painted on the engine cowlings of at least two of the Fokker D IIs. As personal identification for the individual pilots, the planes had differently painted wheel hubs.

Based on the available photos, it seems that not all aircraft of the Fokker-Kommando were painted in grasshopper green. But even if not all of the aircraft were painted in grasshopper-green, this was the first attempt of a "Staffelkennung" for a German fighter unit at the front. The inventor of this locally applied paintwork was Lt. Fritz Grünzweig. He was an artist by profession, and also painted the wild looking faces onto the engine cowlings himself.[27]

In 1974 I visited Gen. a.D. Otto Dessloch who lived in Munich. He was a member of Kampfeinsitzer-Kommando-Ensisheim from the 29 June 1916 to 13 October 1916 when he had to make an emergency landing in Switzerland where he was interned. After his release from internment in Switzerland he was a member of Jagdstaffel 16 from 20 January 1917 to 19 April 1917, when he was appointed to Staffelführer of Jagdstaffel 35. He led the unit until 29 September 1917.

He told me that at the beginning of October 1916 they also started to experiment with camouflage and paint some of their aircraft with rust red, light green and dark green. So, it seems the "grasshopper-green" was not long in use, but it was striking enough that it was well known among the Armee-Abteilung-B.

The Fokker D II 536/16 Lt. Otto Dessloch flew when he had to make a force landing in Switzerland was one of the "experimental camouflaged" Fokker D IIs that had just been painted a few days before.

Lt. Fritz Grünzweig, Fokker D II, Kampfeinsitzer-Kommando, August/September 1916, profile 18

The aircraft had a light "grasshopper" green painted fuselage, and a wild-looking face was painted on its engine cowling. The ground crew used the light green camouflage paint that had recently come into use for application to the upper surfaces of fabric wings as the basis for this "grasshopper green". Paint of this color was available at the local Armee-Flug-Park in substantial quantities.[28] They mixed the light green camouflage paint with a small amount of yellow in order to obtain the desired shade of green. The colors of the face details are not known and were interpreted for the color profile by Alexsandr Kasakov.

Above: Jagdstaffel 16 at Ensisheim airfield regularly received visits from local residents, including many children, who admired the aircraft. At the time, the Elsass (Alsace) region belonged to Germany. Fritz Grünzweig poses here for a photo with children in the pilot's seat of his Albatros D II (LVG).

Fritz Grünzweig was born on 19 September 1892 in Ludwigshafen in the Palatine, which at this time belonged to the Kingdom of Bavaria. He began his service with the Bavarian Fliegertruppe as an observer and joined the Bavarian Feldflieger-Abteilung 9 on 1 June 1915. After being wounded, he trained as a pilot and returned to his old Feldflieger-Abteilung in this capacity on 25 June 1916. On 20

Facing Page: The artist and fighter pilot Lt. Fritz Grünzweig in front of his "grasshopper green" Fokker D II. He was the artist who painted the grim looking faces on the engine cowlings.

Above: Magnification of the "monsters head" painted onto the cowling of his Fokker D II.

August 1916, he was assigned to the Kampfeinsitzer-Kommando at Habsheim. Like a number of other fighter pilots of the Kampfeinsitzer-Kommandos, he was transferred to the newly formed Jagdstaffel 16 on 1 November 1916. Just nine days later, he was transferred to the Bavarian Flieger-Ersatz-Abteilung 1 and assigned as flight instructor to the Bavarian Military Flying School in Schleißheim.

On 15 February 1917, he returned to Jagdstaffel 16. On 14 April 1917, he attacked a French captive balloon. During this sortie he was attacked by French fighter planes and suffered serious wounds in the ensuing combat. He managed to bring down his aircraft in an emergency landing on French territory but died on the way to a hospital.[29]

According to Otto Dessloch he was the "happy heart" of the Staffel, always of good mood.[30]

**Lt. Fritz Grünzweig, Fokker D II,
Kampfeinsitzer-Kommando,
August/September 1916, profile 18**

3.2 Summary

The available photo material of the German planes of the years 1914–1916 shows that a number of variants of identifications emerged then that would resurface later on the single-seaters operated by the Jagdstaffeln. These were, for example, numbers and symbols, the colors of the cap bands or the colors of the peace uniforms of the parent units of the pilots. Other examples were the coats of arms, especially in the case of pilots that came from noble families, the first letter of the surname and various other symbols, such as names or animals.

Individual indentifiers were applied to aircraft in an increasing manner as the year 1916 progressed. But the application of individual markings, let alone the use of Staffel markings, were still the exception. There is evidence of some kind of identification in the form of numbers or symbols in some flying units, but this was not a Staffel identification in the true sense, but rather an ordering system for the deployment of the aircraft assigned to each aviation unit. Kampfgeschwader I, IV and VI, for example, used a system of numbers for identification, while Kampfgeschwader II used, according to available material, mainly different symbols.

A comparison of the aircraft markings of the Kampfgeschwader with the markings of the aircraft of the first Jagdstaffeln shows that those markings became a role model for the markings of the aircraft in several of the first Jagdstaffeln.

In contrast, the black-and-white or black-white-red bands used several times by the Fokker fighters were hardly used at first by the new Jagdstaffeln. The new German fighters, especially the Albatros D I and D II, differed drastically in silhouette from all Allied fighters then flying at the front, so this identification was no longer necessary.

4. The Formation of the First Ten German Jagdstaffeln in 1916

At the beginning of the Battle of the Somme in July 1916, the fighting strength of the German air force had reached its lowest level, as previously described. Allied fighter units, protected by the new British Airco D.H. 2 fighters, dominated the airspace above the battlefield. Even the new British two-seaters such as the F.E. 2b were superior to the German Fokker E IIIs and E IVs, not to mention the Pfalz single seaters. If German airplanes appeared over the lines, they became an easy prey for the new British aircraft. Without effective protection from their own fighter pilots, the German airplanes of the Artillerieflieger-Abteilungen, Feldflieger-Abteilungen and Kampfgeschwader were extremely limited in their ability to carry out missions. The ground troops felt helpless and abandoned by their own Fliegertruppe.

In an effort to rectify this situation, the first German Jagdstaffeln were formed in August 1916. These new fighter units were permanent formations of aircraft and pilots that were intended to function as a cohesive unit. On 31 August 1916, the order was given to form 20 Jagdstaffeln until the end of the year. Out of this total, the formation of the first

five Jagdstaffeln had to be completed before the end of August 1916. Two were to be assigned to the German 1st Army, two to the 2nd Army, and the remaining one to the 5th Army.[1]

When the Staffelführer of these first Jagdstaffeln gathered their pilots around them, it turned out that the airmen assigned to them had quite different skills and frontline experience.

A number of the pilots came from the Kampfeinsitzer-Kommandos. They had flown Fokker and Pfalz rotary engine monoplanes for an extended period. Some had experience with the Fokker D II biplane equipped with a rotary engine and only a few had experiences with the new Fokker, Halberstadt and Albatros biplanes that were powered by inline engines and therefore had different flight characteristics.

These pilots knew how to attack enemy aircraft and had gained extensive air combat experience in previous missions. However, thus far the pilots of the Kampfeinsitzer-Kommandos had largely flown their missions alone. Flying in close formation and attacking the enemy as a cohesive unit was a new and unknown challenge for them. As it turned out,

Above: Oswald Boelcke's four mechanics attending his Fokker E IV 123/15 at Sivry. The 160 hp from the twin-row Oberursel U III engine made the E IV notably faster than earlier E-types with 100 hp engines. The E IV had similar speed to the Nieuport 11 and DH.2 and the ability to mount two synchronized guns, but both Allied fighters had better climb and maneuverability. (photo courtesy of Charles G. Thomas)

Right: The Pfalz E IV used the same 160 hp twin-row Oberursel U III engine as the Fokker E IV. Again its speed in level flight was similar to the Nieuport 11 and DH.2, but both Allied fighters had better climb and maneuverability. Most Fokker and Pfalz monoplanes used less powerful engines, which made them slower. Both types were developed from pre-war airframes too fragile for high-G maneuvering combat; both the DH.2 and Nieuport 11 biplanes were more robust and withstood such combat much better.

even some experienced Kampfeinsitzer pilots had problems adjusting to the new method of flying and fighting.

Moreover, with the appearance of the Allied fighter planes of the Nieuport XI and Airco D.H. 2 type, the manner of aerial combat had also changed.

The maneuverability of the new Allied fighter planes forced the German fighter pilots increasingly to fight in tight turns, a fighting style that was unfamiliar to them since their Fokker and Pfalz monoplanes were simply not suitable for this due to their flight characteristics. The wing-warping mechanism of the Fokker monoplanes resulted in relatively limited maneuverability compared to the newer types fielded by the opposing side, which all employed aileron control.

Some of the former Fokker fighter pilots succeeded in adapting to the new fighting style, but in other cases even experienced Kampfeinsitzer pilots could not cope with this new style of combat in the air. Usually, they were gradually transferred to the rear areas, where they could pass on their experience to young student pilots as flight instructors.

Another portion of the pilots came from the Kampfstaffeln of the Kampfgeschwader. The number of pilots of these units in the Jagdstaffeln increased considerably at the end of 1916, after the disbanding of the Kampfstaffeln. A number of the Kampfstaffeln were equipped with twin-engine bombers to form the Bomberstaffeln in 1917. Other Kampfstaffeln were converted into Schutzstaffeln and retained their two-seaters. Their task was to protect the artillery cooperation aircraft of the Fliegerabteilungen (A) against attacks by hostile aircraft.

Johann Janzen, a member of Kampfstaffel 12 of Kampfgeschwader II, told Alex Imrie that the pilots of the Kampfstaffeln considered themselves as the elite of the Fliegertruppe and had little interest in flying the heavy twin-engine bombers or in circling over the front lines and playing "governess" for the artillery cooperation aircraft. A deployment as a fighter pilot seemed much more exciting and thrilling to them. Accordingly, he enlisted in the newly formed Jagdstaffeln, as did a good number of his comrades [2] The pilots of the Kampfstaffeln were experienced in flying their missions in close formation but lacked experience in air combat. They had flown two-seater aircraft and now had to learn to fly the fast and maneuverable single-seat fighters. An exception was the small group of Fokker fighter pilots assigned to the Kampfstaffeln as protection, such as Hans Bethge, who was transferred to Jagdstaffel 1 and later became Staffelführer of Jagdstaffel 30.

Finally, airmen from the Feldflieger-Abteilungen and Artillerieflieger-Abteilungen arrived at the Jagdstaffeln. They had also flown two-seaters and now also had to learn to fly the fast and nimble new fighter planes. Moreover, they had flown most of their missions alone. Their task so far had been to avoid combat with enemy aircraft, because only in this way they could fulfill their mission of reconnoitering and documenting the enemy's

Above: The Nieuport 11 sesquiplane was more robust than the Fokker and Pfalz monoplanes and generally out-performed and out-maneuvered them. A fixed machine gun was mounted on top of the wing to fire over the propeller arc. The Nieuport 11 was the first of an entire family of Nieuport fighters with more power and some with a synchronized gun for the pilot. Here leading French ace Jean Navarre is patrolling over Verdun.

movements or directing their own artillery fire. Thus, they had little experience in both aerial combat and flying in close formation.

The task of the Staffelführer of the newly formed Jagdstaffeln was to form a fighting unit from pilots with different experiences and different skills. They had to train them to be able to attack as close as possible while maintaining formation, with coordinated flight maneuvers at the right moment and from the right position. **But no matter what unit they came from or what flying background they had, all these young pilots were filled with great enthusiasm for their new military assignment, as Oswald Boelcke, Staffelführer of Jagdstaffel 2 reported:**

"My gentlemen are all very passionate about it and very eager and capable, but I have yet to train them properly to work together calmly - now they are sometimes still like young dachshunds in their eagerness to accomplish something."[3]

When the first Jagdstaffeln entered the air war in August 1916, there were no guidelines issued by higher authorities concerning their actual deployment at the front. Their mission was simply formulated: "To gain aerial supremacy". This meant to enable their own aircraft to fulfill their missions

such as reconnaissance, artillery cooperation and bombing and to prevent the enemy air forces from carrying their missions out over German-held territory.

Thus, it was left to the first Staffelführer to decide how their mission was to be interpreted and practically implemented. More than at any other time, the success of the Jagdstaffel depended on the ability and performance of these first Staffelführers. It was mainly thanks to the written experiences of these men that the first directives for the deployment of the new Jagdstaffeln could be issued at all. These are inseparably linked to the achievements and experience reports of Staffel leaders such as Hptm. Martin Zander, Hptm. Oswald Boelcke, Oblt. Hans Berr, Oblt. Hans-Joachim Buddecke and Oblt. Kurt Student.

The field reports written by this handful of officers formed the nucleus for the "Weisungen für den Einsatz und die Verwendung von Fliegerverbände innerhalb einer Armee "(Instructions for the Deployment and Use of Aviation Units Within an Army). This was issued by the Chief of the General Staff of the Field Army, published in May 1917, and, among other things, it formulated the intended use of the Jagdstaffeln:

Above: The DH.2 pusher biplane was more robust than the Fokker and Pfalz monoplanes and generally out-performed and out-maneuvered them. Although high drag, which precluded future development, its pusher layout enabled a single fixed machine gun firing forward for the pilot.

Point: 37: *"The battle for air supremacy must always be waged offensively. The enemy aircraft have to be sought out behind their own lines, attacked and, whenever they advance beyond our lines, driven back."*

Point 40: *"The only effective weapon of the struggle for the enforcement and assertion of air supremacy are the Jagdstaffeln."*[4]

The missions of the Jagdstaffeln were finally specified in the "Weisungen über den Einsatz von Jagdstaffeln" (Instructions on the Deployment of Jagdstaffeln), published in October 1917 by the Chief of the General Staff of the Field Army. The following excerpted contents provide a good overview of the tasks of the German Jagdstaffeln:

Point 1: *Fighting an enemy airplane is, next to balloon attacks, the most important task of the fighter pilot; and to seek the enemy out, his noblest duty. Only the loss caused by the shooting down of numerous hostile aircraft imposes the feeling of superiority on the enemy.*

Point 2: *The fighter pilot therefore finds his success in attack only.*

Point 5: *The elimination of enemy aerial and balloon observation is of decisive importance for*

the artillery battle; therefore, the main task of the Jagdstaffeln is to eliminate the planes that guide the fire of the artillery behind their own lines by repeated advances across the front line to taking away the eyes of the enemy artillery.

Point 7: *Enemy reconnaissance aircraft and bomb squadrons flying over the front at high altitude can often no longer be caught in time by fighter pilots who only take off when informed of their appearance. It is therefore necessary to send individual Jagdstaffeln to hunt freely against them."*[5]

This "free hunt" that is being referred to in the last sentence meant that the aircraft of the Jagdstaffeln flew to the front on the basis of their estimation of the air combat situation and the instructions of the Staffelführer to look out for and attack Allied aircraft. In this case, the initiative regarding these missions lay in the hands of each Staffelführer.

The nominal strength of a Jagdstaffel consisted of 12 pilots and 14 aircraft, but the first Jagdstaffeln hardly reached this strength. Thus Hptm. Oswald Boelcke wrote in his letter of 17 September 1916 from Bertincourt airfield:

"The Staffel is still not quite finished, because I am still missing half of the machines. At least six arrived yesterday - so I can take off with my Staffel for the first time today. I've flown Fokker biplanes so far, but today I want to take one of the new Albatros."[6]

4.1 Royal Prussian Jagdstaffel 1

The majority of the Staffelführers of the Jagdstaffeln formed in 1916 such as Hptm. Oswald Boelcke, Oblt. Hans-Joachim Buddecke, Oblt. Hans Berr, Oblt. Fritz von Bronsart-Schellendorf or Oblt. Kurt Student, were able to form and lead a powerful, successfully operating unit. Other Staffelführers, for all their personal courage and experience as front-line pilots, had problems doing so. They were eventually transferred as leaders to other units, such as the Flieger-Abteilungen (A), to flight schools or to the Jagdstaffel schools which were soon to be formed, where they usually rendered excellent service. On August 22, 1916, in accordance with the order of 10 August 1916, Jagdstaffel 1 was established in the German 1st Army area.[1] Its first airfield was Bertincourt, which was used until 23 September 1916, after which the Staffel moved to Hermies.[2]

The nucleus of the Staffel was provided by Kampfeinsitzer-Kommando Nord. Additional ground personnel and material were taken over from Armee-Flug-Park 1 and from various Feldflieger-Abteilungen of the 1st Army. Hptm. Martin Zander, until then leader of Kampfeinsitzer-Kommando Nord, was appointed as the Staffelführer of the new unit. According to available photos and documents, the early equipment of the Staffel consisted of Halberstadt D II, Fokker D I and Albatros D I aircraft.[3]

This mix of aircraft types seems surprising at first glance. However, it must be remembered that in August the new fighter biplanes of these three companies were still quite new and only available in relatively small numbers. Equipping the first Jagdstaffeln with a single type of aircraft was not yet practical, and the same was true of the other early Jagdstaffeln that would soon follow. Furthermore, the "performance race" between the various new fighter types had not yet been decided. The Halberstadt D II and D III were the first to reach the frontline units, followed by the Fokker D I to D III and soon after the Albatros D I arrived. These types were given a fair chance to prove their merits, and for a few months the outcome of this comparison was not clear.

On 24 August 1916, Offz. Stellv. Leopold Reimann scored the Staffel's first aerial victory. He attacked an English aircraft referred to as a Sopwith east of Metzen-Coutre and shot it down at about 18:30. This was also the first aerial victory of a German Jagdstaffel. The next day, Hptm. Martin Zander was successful. He forced a plane of the 22nd Sqn RFC designated as a "Vickers" to land at Gueudecourt inside German lines. The crew was captured. Four days later, Oblt.

Above: Sanke postcard of Hptm. Martin Zander, who served as the first Staffelführer of Jagdstaffel 1 from 22 August 1916 to 10 November 1916. He became the head of the newly established Jagdstaffelschule in Valenciennes afterwards.

Hans Bethge reported his first aerial victory: a British aircraft identified as a B.E., crashed at 12:05 under his fire near Auchonvillers.[4]

The last day of the month, 31 August 1916, was the first day of good weather after a rainy week. Accordingly, increased flying activity was observed on both sides of the front on that day. In the morning at about 8:00, aircraft of Jagdstaffel 1 and Kampfgeschwader I attacked a group of Martinsyde

Above: Two early-production Fokker D Is of the Jagdstaffel in front of the airplane tents at Bertincourt airfield in August/September 1916. (L. Bronnenkant).

Above: Lt. Kurt Wintgens was one of the first successful fighter pilots of Jagdstaffel 1. He achieved his 14th to 18th aerial victory while flying with this unit. The photo shows him in front of his light blue Halberstadt D II in June 1916, prior to the formation of Jagdstaffel 1.

G. 100s of 27th Sqn. R.F.C. Lt. Hans von Keudell, Oblt. Hans Bethge and Lt. Gustav Leffers each scored an aerial victory. A fourth aircraft was shot down by the crew of Lt. Willi Fahlbusch and Lt. Hans Rosencrantz of Kampfstaffel 1.[5]

Left: Pilots of Jagdstaffel 1 at Bertincourt airfield in September 1916. From left: Lt. Reifhäuser, Offz.Stellv. Karl Ehrnthaler, Lt. Gustav Leffers, Lt. Goede, Offz. Stellv. Teka and Staffelführer Hptm. Martin Zander. (A. Imrie)

Below: Albatros D I D.385/16 just after being newly delivered to the Staffel. This was one of the prototype Albatros D I aircraft, and it likely was the first example of this type to reach the front. The water cooler mounted in front of the engine was only mounted to two or three of the D I prototypes. All indications are that it was this Albatros D I that Vzfw. Reimann brought along when he was transferred to Jagdstaffel 2 on 1 September 1916. Note the mottled camouflage on the rudder, which was a feature noted on several prototypes of the Albatros D-type fighters. (A. Imrie)

Right: Sanke card of Offz. Stellv. Leopold Reimann, who achieved the first aerial victory of a German Jagdstaffel. On 24 January 1917, he crashed fatally during a training flight over the airfield at Valenciennes.

Unser erfolgreicher Kampf-Flieger
Offz.-Stellv. Reimann

429
Postkartenvertrieb W.Sanke
BERLIN N.37.
Nachdruck wird gerichtlich verfo

4.2 Royal Prussian Jagdstaffel 2

On 10 August 1916, instructions were issued to form Jagdstaffel 2 in the area of the German 1st Army.[1] Hptm. Oswald Boelcke, who was on a visit to his brother Wilhelm at Kampfgeschwader II in Kovel on the Eastern Front, was ordered back to the Western Front on 14 August 1916. Under his command, Jagdstaffel 2 was assembled on 27 August 1916. The Staffel was assigned to Vélu, about 2 km north of Bertincourt in the German 1st Army area, as its airfield. The transcript of the war diary recorded the following entry:

Inventory: 3 officers, these are: Hptm. Boelcke, Lt. von Arnim and Lt. Günther, 64 NCOs and enlisted men, the airfield is Vélu, where 4 hangars are taken over from Fliegerabteilung 32. Accommodations for officers are provided at Bertincourt. Aircraft are not yet available. The first planes arrived on September

Right: Hptm. Oswald Boelcke, the master teacher of the German Jagdstaffeln.

Facing Page: Hptm. Oswald Boelcke in the pilot's seat of his Fokker D III 352/16. Note the oil stains on the fuselage fabric.

Below: Hptm. Oswald Boelcke in front of his Fokker D III 352/16 with which he achieved the first aerial victory credited to Jagdstaffel 2 on 2 September 1916.

Above: Lt. Erwin Böhme in front of his Fokker D I 185/16 at Vélu airfield. Apparently this aircraft, along with Fokker D III 352/16, arrived at Jagdstaffel 2 on 1 September 1916. These were the first two aircraft delivered to the newly-formed Jagdstaffel.

1: Vfw. Reimann brought one Albatros D I along from Jagdstaffel 1 and a Fokker D III was picked up from the Armee-Flug-Park.[2]

Besides this Fokker D III, photographic evidence indicates that Fokker D I 185/16 also arrived at the unit on this date.

On September 2, 1916, Hptm. Oswald Boelcke took off on a front-line flight in his newly-arrived Fokker D III 352/16. Northeast of Thiépval he

attacked a British aircraft that he identified as a B.E. In the course of this action, he in turn was attacked by three D.H. 2s, whereupon Hptm. Boelcke had to break off the attack. One of the D.H. 2 pilots, Capt. R. Wilson of 32 Sqn. RFC, pursued him across the lines. Oswald Boelcke and Capt. Wilson soon got involved in a short but fierce dogfight during which the D.H. 2 received numerous hits and was forced to land. The British pilot managed to jump out of

Above: The Albatros D I, (presumably D.385/16) fitted with the conspicuous water tank in front of the engine on the field airfield Vélu near Bertincourt. To the right of it stands Fokker D III 352/16, which was the preferred mount of Oswald Boelcke prior to the arrival of his Albatros D II D.386/16, on 16 September. (L. Bronnenkant)

his machine, which caught fire on impact, and rid himself of his coat, which was also on fire. He spent the following day at Bertincourt airfield as the guest of Hptm. Boelcke and his men. This was the first aerial victory of Jagdstaffel 2.

Six days later, on 8 September 1916, Hptm. Boelcke gained the Staffel's second aerial victory. He again took off alone and attacked a D.H. 2 from 24 Sqn. RFC diagonally from behind. After taking several bursts of fire his opponent began to burn, exploded shortly thereafter, and went down spinning like a top, the pilot falling out in the process.

The following day, Hptm. Oswald Boelcke again took off alone toward the front. Southwest of Bapume he sighted four D.H. 2 from 24 Sqn. circling above the lines. He attacked one of the D.H. 2s which caught fire after a short burst. The British pilot, Lt. N. P. Mansfield,

attempted to glide over the lines, but there the plane crashed, killing him.[3]

In the first weeks of its existence, Jagstaffel 2 was shaped like no other Jagdstaffel by the leadership personality of its Staffelführer, Oswald Boelcke, as Erwin Böhme described in his letter dated 18 October 1916:

"There is something quite special about the way Boelcke transfers his spirit to each and every one of his students, the way he carries them all along. They go with him wherever he leads them, none would ever abandon him. He is an absolute leader!

No wonder his Staffel is flourishing. We have now received a number of excellent young people as additions to the Staffel. The victories of the Staffel are piling up. Yet, despite the many battles and many a bold venture, we have not suffered a single loss in the last two weeks."[4]

4.3 Royal Prussian Jagdstaffel 3

In contrast to the two previously mentioned Staffeln, Jagdstaffel 3 was formed in the homeland. According to the order dated August 1916, the formation took place at Flieger-Ersatz-Abteilung 5 in Braunschweig.[1] Oblt. Kohze, who was transferred to the Staffel from Kampfgeschwader IV, was appointed Staffelführer. On 30 August 1916, Vfw. Miesegades was killed in a test flight at the controls of a Fokker D I at Flex airfield.

Two days later, on 1 September 1916, the Staffel was mobilized and began transferring to the German 2nd Army, where it moved to Vraignes Airfield, where it would remain until 5 November. According to the war diary transcript, the Staffel was equipped at the front with Halberstadt D III aircraft.[2]

However, photos of the Staffel with Halberstadt D III aircraft are not available. The first available photos of the Staffel show it being equipped with Albatros D IIs in late 1916, likely around November, at Fontaine-Uterte airfield. These photos indicate that Jagdstaffel 3 may have received Albatros D IIs as early as late October/early November, since several of the fighters shown are D IIs from the first production batch, which reached the front around that time.

As early as 12 September 1916, the Staffel suffered its first air combat loss when Lt. Ewald von Mellenthin fatally crashed after air combat beyond the lines at Poziéres.[3]

On 15 September 1916, Lt. Mohr, Lt. von Massow, and Lt. von Busse attacked an Allied aircraft identified as a Sopwith two-seater south of Péronne, which landed on the German side. However, the aerial victory was awarded to Lt. Wilhelm Frankl

Above: Oblt. Kohze the Staffelführer of Jagdstaffel 3.

of Jagdstaffel 4, who also took part in the fight. The Staffel's first confirmed aerial victory was achieved by Lt. König. He shot down an unspecified enemy aircraft north of Bouchavesnes on 22 September 1916.[4]

4.4 Royal Prussian Jagdstaffel 4

The Jagdstaffel was established on 25 August 1916 at Roupy airfield, in the area of the German 2nd Army.[1] The nucleus of the unit was provided by Kampfeinsitzer-Kommando Vaux, also named Kampfeinsitzer-Kommando Süd. Additional personnel and equipment came from Armee-Flug-Park 2 and the Army's Feldflieger-Abteilungen. Oblt. Hans-Joachim Buddecke was appointed Staffelführer.[2] According to available sources, the Staffel was initially equipped with Halberstadt D-Type fighter airplanes, at least two Fokker D Is, and some remaining Fokker E monoplanes.

On 6 September 1916, Oblt. Hans-Joachim Buddecke, Lt. Otto Bernert and Oblt. Rudolf Berthold attacked a Caudron formation near Dompierre, east of Peronne. In the aerial fight, Lt. Bernert succeeded in bringing down a Caudron of Escadrille C 28, thus scoring the first aerial victory for the Staffel.[3] The very next day Lt. Wilhelm Frankl was successful. He was able to bring down a Nieuport of Escadrille N 73 after aerial combat northeast of Combles.[4]

Right: Oblt. Hans-Joachim Buddecke, the first Staffelführer of Jagdstaffel 4, wearing a Turkish uniform from his service as a Turkish officer at the Dardanelles front.

Below: Jagdstaffel 4 shortly after its formation at Roupy airfield. In the picture four Halberstadt single-seaters, two Fokker D Is and one dismantled Fokker E IV can be made out.

Above: Oblt. Hans-Joachim Buddecke in the pilot's seat of his Halberstadt D V, autumn 1916. The compass mounting in the left lower wing is rarely seen in photos of this type of aircraft.

Left: The start-house of Jagdstaffel 4 is under construction at Vaux airfield in September 1916. Behind it, the tails of two Halberstadt D-type fighters of the Staffel can be seen.

4.5 Royal Prussian Jagdstaffel 5

On 21 August 1916, Fokkerstaffel-Avillers was formed into Jagdstaffel 5 in the German 5th Army area.[1] The leader of the Fokkerstaffel, Oblt. Hans Berr, was appointed Staffelführer. The Staffel was allocated an airfield near Beauchamp, east of Etain, on the same day. Initially, Fokker E III and Fokker E IV aircraft were available, soon to be replaced by Halberstadt D II/D V.[2]

Already on 26 August 1916, the Staffel's first aerial victory was achieved. Sources differ on who scored this victory. According to the compilation of the victories of the unit by Dr. Gustav Bock, it was Vzfw. Paul Piechl who shot down a Voisin west of Moulainville on 26 August 1916. Other documents name Vfzw. Hans Müller as the victor in this air fight.[3]

In any case Vzfw. Müller scored the second victory for the Staffel on 31 August 1916 when he shot down a captive balloon southwest of the Maasbogen (bend of the Meuse River).

On 26 September 1916, the Staffel was first transferred to Bellevue-Ferme airfield near Senon and three days later to Gonnelieu in the German 1st Army section, where it was stationed until March 1917.[4]

Right: Oblt. Hans Berr, the popular first Staffelführer of Jagdstaffel 5.

Below: The first known group picture of Jagdstaffel 5, taken around October 1916. Seen from left: 1. Lt. Max Böhme, 2. Lt. Kutscher, 3. Lt. Schneider, 4. Lt. Hans Gutermuth. 5. (unknown), 6. (unknown), 7. Vfw. Paul Hoppe. 8. (unknown) 9. Vzflgmstr. Max Winkelmann, 10. Offz.Stellv. Edmund Nathanael, 11. (unknown), 12. (unknown), 13. Lt. Renatus Theiller.

Above: Halberstadt D V in front of one of the aircraft tents on the airfield at Gonnelieu in the late autumn of 1916.

Left: Vzfw. Hans Müller probably achieved the first aerial victory of Jagdstaffel 5. The photo shows him as a member of Fokkerstaffel Avillers in front of his Fokker monoplane baptized *Frechdachs* (very commonly used German admonition with the connotation of a "smart-alec"). (G. VanWyngarden)

4.6 Royal Prussian Jagdstaffel 6

According to the order of the Feldflugchef, Fokkerstaffel-Sivry was disbanded on 22 August 1916 and converted into Jagdstaffel 6.[1] The Staffelführer was Rittm. Josef Wulff, who arrived at the Staffel two days later from Kampfgeschwader V. The airfield was Sivry in the German 5th Army area. A war diary entry dated 2 September 1916 notes that aerial combats of the Jagdstaffel had been unsuccessful up to this point.[2]

Among the first airmen assigned to the Staffel was Vzfw. Carl Holler, who was transferred from Fokkerstaffel-Sivry on 25 August 1916.

In his book "*Als Sängerflieger im Weltkrieg*" he wrote:

"Besides our eight Fokker machines, the Jagdstaffel had a two-seater plane. It served to show the artillery officers from the air how far they had succeeded in covering their positions against observation by aviators."[3]

On 23 September 1916, Sivry airfield came under fire from long-range enemy artillery. The Staffel departed their base and transferred to Jametz airfield. Six days later, the Staffel was transferred to the German 2nd Army, where it occupied Ugny le Quipée airfield. On 20 October 1916, Vzfw. Kress finally scored the first aerial victory for the Staffel.

Above: Rittm. Josef Wulff, the first Staffelführer of Jagdstaffel 6.

Below: Briefing at the airfield of Jagdstaffel 6, from left: 1. Lt. Deilmann, 2. Werkmeister (foreman of the mechanics), 3. Rittm. Wulff. (W. Bock)

Above: Albatros D I D.457/16 serves as the backdrop for members of Jagdstaffel 6. Left to right: 1. Vzfw. Holler, 2. (unknown) 3. (unknown), 4. Vzfw. Fritz Loerzer, 5. (unknown). (G. van Wyngarden).

Southwest of Peronne, beyond the lines, he attacked an aircraft identified as a Morane and shot it down. Six days later, Lt. Nauck was successful. His opponent was an aircraft identified as Farman, which crashed near Omiecourt on the German side.[4]

4.7 Royal Prussian Jagdstaffel 7

On 22 August 1916, the formation of Jagdstaffel 7 began at Martincourt airfield, in the area of the German 5th Army.[1] The nucleus of the new Jagdstaffel was the "Fokkerstaffel bei der 34. Infanterie Division" (Fokkerstaffel located at the 34th Infantry Division). The previous leader of the Fokkerstaffel, Oblt. Fritz von Bronsart und Schellendorf, was appointed Staffelführer. Fokker E III and E IV aircraft were available as initial equipment of the new unit. On 29 September, the Staffel moved to Bellevue-Ferme, east of Senon.[2]

It is not known when the unit converted to the new biplane fighters. On 13 October 1916, Lt. Wilhelm Eckenberg was transferred to the Staffel and belonged to it until 14 July 1917. A photo in his album shows the Albatros D II he flew with Jagdstaffel 7. An exact date for when this photo was taken is not given.[3]

It took until 23 October 1916 for Lt. Hermann Kunz to finally achieve the first aerial victory of Jagdstaffel 7. Flying alongside fellow Staffel member Vzfw. Schramm, he attacked two Nieuports over Maucourt, where he succeeded in shooting down his opponent in a dogfight. The Staffel's first aerial victory was credited to Lt. Hermann Kunz. On 2 November 1916, the Staffel was transferred to La Folie Ferme. On 11 November 1916, Lt. Kunz was again successful. In aerial combat with several Nieuports, he succeeded in bringing one of them down inside the German lines.[4]

Above: Oblt. Fritz von Bronsart und Schellendorf, the first Staffelführer of Jagdstaffel 7, conveniently obscured the military number of his Fokker E IV.

4.8 Royal Prussian Jagdstaffel 8

The order for the formation of Jagdstaffel 8 was issued on 10 September 1916.[1] The formation of the Staffel began two days later. Feldflieger-Abteilung 6 provided five aircraft, Artillerieflieger-Abteilung 213 provided two aircraft with ground crew, and Feldflieger-Abteilung 40 and Feldflieger-Abteilung 33 each provided one aircraft with ground crew. On 23 September 1916, Hptm. Gustav Stenzel, previously leader of Kampfstaffel Vardar at the Macedonian front, arrived as Staffelführer.

The unit's first airfield was Rumbeke in the German 4th Army area. According to the research of Greg VanWyngarden and photos in his collection, the Staffel was equipped with Fokker D IIs and at least one Albatros D I after its formation. Subsequently, the Staffel converted to Albatros D IIs.

As early as 1 October 1916, Uffz. Ulmer scored the first aerial victory by shooting down a captive balloon at Oostvleteren.[2]

Above: Hptm. Gustav Stenzel, the first Staffelführer of Jagdstaffel 8.

Right: Vzfw. Alfred Ulmer achieved the first aerial victory of Jagdstaffel 8. He was killed in aerial combat on 29 June 1917. (W. Bock)

Below: Members of Jagdstaffel 8, from left: 1. Lt. Philipp Wieland, 2. (unknown), 3. Hptm. Gustav Stenzel, 4. (unknown). (W. Bock)

4.9 Royal Prussian Jagdstaffel 9

In February 1916, the Fokker single-seat fighters serving with Feldflieger-Abteilungen of the German 3rd Army had been combined into two Fokkerstaffeln. These were Fokkerstaffel A at Leffincourt airfield and Fokkerstaffel B at Merland-Ferme airfield, near Pont Faverger. Both Fokkerstaffeln were combined on 1 June 1916 to form the "Armeestaffel des Armee-Ober-Kommando (A.O.K.) 3" (Fokkerstaffel of the Army Headquarters of the 3rd Army) at Vouziers airfield. Leader of the Fokkerstaffel was Oblt. Kurt Student. On 1 July 1916, the Staffel moved to Leffincourt airfield.[1]

On 28 September 1916, the order was given to establish Jagdstaffel 9. The day of mobilization of the Jagdstaffel was 5 October 1916. By this time, A.O.K. Staffel 3 had already gained 15 aerial victories, making it one of the most successful single-seat fighter units at the Western Front.[2]

In the fall of 1975, I visited retired General Kurt Student in Lemgo in northern Germany. Despite the fact that he was working intensively on his book about the German paratroopers in World War II, he took time for an interview, and I was able to photograph his extensive photo album. According to him, Jagdstaffel 9 initially took over Fokker E III and E IV aircraft from the A.O.K. Staffel, but they were soon supplemented by Fokker D II fighters and later Halberstadt fighters were delivered to the unit.

As General Student further explained in the interview, when the Fokkerstaffel was transformed into Jagdstaffel 9, he had the great advantage that his Staffel consisted of experienced fighter pilots who had known each other for months. When he had been the leader of A.O.K.-Staffel 3, he had regularly ordered his pilots to fly in groups of two or three to the front. Operations in formations of six or more aircraft were therefore no great challenge for them.[3]

On 22 October 1916, Lt. Adolf Frey gained the unit's first aerial victory. He attacked a Nieuport single-seater at Tahure and a cornering battle fought with all ferocity ensued. In the dogfight, the Nieuport came under machine gun fire from the German aircraft and crashed beyond the lines west of Arbre Heights near Tahure.

Above: Oblt. Kurt Student, leader of "Armeestaffel des Armee-Ober-Kommando (A.O.K.) 3" and Jagdstaffel 9.

On 3 November 1916, Lt. Hartmuth Baldamus reported shooting down a Nieuport beyond the lines near Tahure. However, since the impact of the enemy aircraft could not be observed from the German side, the claim was not granted confirmation. On 9 November 1916, the Staffel received its first Halberstadt D IIs. On the same day, Lt. Baldamus attacked a Nieuport and brought it down. This time the aerial victory was confirmed as the Staffel's second aerial victory and Lt. Baldamus' sixth kill.[4]

Above: The pilots, crews, and aircraft of the "Armeestaffel des Armee-Ober-Kommando (A.O.K.) 3" formed the nucleus of Jagdstaffel 9, seated second row the pilots from left: 1. Lt. Schlolaut, 2. Lt. Dinkel, 3. Oblt. Student, 4. Oblt. Zietlow, 5. Lt. Lehmann, 6. Vfw. Koehler.

Below: As a replacement for the Fokker monoplanes, the Staffel received Fokker D IIs and at least one D III. The photo shows Lt. Hermann Pfeiffer's Fokker D III with his mechanics. Note the mud-guards fitted above the wheels. (M. Schmeelke)

Above: Lt. Adolf Frey, seen to the right of the later "Pour le Mérite" holder Erich Loewenhardt, achieved the first aerial victory of Jagdstaffel 9 on 22 October 1916. Lt. Frey was killed in aerial combat on the last day of "Bloody April" 1917.

Left: Following equipment with the Fokker D-types, the squadron also received Halberstadt single-seaters. The photo shows Lt. Werner Marwitz in front of his Halberstadt D II at Leffincourt airfield.

4.10 Royal Prussian Jagdstaffel 10

On 21 September 1916, the German 6th Army issued a daily order to establish a Jagdstaffel in their area.[1] Oblt. Ludwig Linck from Feldflieger-Abteilung 18 was appointed Staffelführer. Four days later, on 25 September 1916, "Jagdstaffel Linck" was established with the Army's Prussian Fokker fighter pilots at Phalempin airfield. The equipment of the Staffel at this time consisted of 4 Fokker E IIIs, 1 Fokker E IV, 2 Fokker D IIs, 1 Halberstadt D II, and 2 Albatros D Is. On 6 October 1916, the Jagdstaffel received the official designation Jagdstaffel 10.[2]

On 22 October 1916, the Staffel suffered its first loss. The Staffelführer, Oblt. Ludwig Linck was shot down in flames after aerial combat with an F.E. 8 over Carvin. He was flying an Albatros D II.[3] Five days later, the Staffel transferred to the German 5th Army and moved to Jametz airfield. On 6 November 1916, Oblt. Helmuth Volkmann, previously of Jagdstaffel 6, was transferred to the Staffel as its new Staffelführer.

It took until 10 December 1916 for the Staffel to be credited with its first aerial victories. On that day, northwest of Verdun, there was a fierce dogfight between aircraft of Jagdstaffel 10 and French Nieuports. According to German 5th Army records, Lt. Gustav Nernst and Lt. Bordfeld each shot down one of the enemy planes during that fight. Additionally, Offz.Stellv. Hebbe reported that he had shot down two enemy planes identified as Nieuports. All aerial victories were confirmed and Jagdstaffel 10 was cited in the daily report of the 5th Army.

Above: The first Staffelführer of the Jagdstaffel 10, Oblt. Ludwig Linck. He was killed in action as early as 22 October 1916.

Above: Members of Jagdstaffel 10 assembled for a group shot in front of two of the squadron's Albatros D I fighters at Phalempin airfield in September/October 1916. From left: 1. Vfw. Barth, 2. Lt. Eugen Weber, 3. Lt. von Fabrice, 4. Lt. Dr. Weber, 5. Lt. Ernst Hess, 6. Vzfw. Rudolf Heinemann, 7. Flugmeister Schnell, 8. Lt. Offermann (?), 9. Lt. Gustav Nernst, 10. Lt. Walter Bordfeld.

4.11 Summary

By the end of August 1916, 25 Halberstadt biplane fighters, 29 Fokker biplane fighters and a single example each of the Albatros D I and D II were in service on the Western Front.[1] Due to the gradual formation of the Jagdstaffeln, one to three Jagdstaffeln had to cover the front area of a German army in August-September 1916 until further newly formed Jagdstaffeln arrived at the front and were operational.

According to the listing in the "**Übersicht** der Behörden und Truppen in der Kriegsformation, Part 10 Luftwaffe – Section B: Fighter Jagdstaffel Formationen" (Overview of the Authorities and Troops in the War Formation, Part 10 Luftwaffe – Section B: Fighter Jagdstaffel Formations), orders had been issued to establish a total of 32 Jagdstaffeln by the end of 1916. These were Jagdstaffel 1 to 26, and 28 to 33. The order for the formation of Jagdstaffel 27 was dated 5 February 1917.[2]

First Deployment of the First 10 German Jagdstaffeln		
German Army	**Jagdstaffel**	**Airfield**
1. Army	Jagdstaffel 1	Bertincourt
1. Army	Jagdstaffel 2	Vélu
2. Army	Jagdstaffel 3	Vraignes
2. Army	Jagdstaffel 4	Roupy
5. Army	Jagdstaffel 5	Avillers
5. Army	Jagdstaffel 6	Sivry
5. Army	Jagdstaffel 7	Martincourt
4. Army	Jagdstaffel 8	Rumbeke
3. Army	Jagdstaffel 9	Leffincourt.
6. Army	Jagdstaffel 10	Phalempin

5. The Military Reasons for the Colorful Painting of the German Jagdstaffeln

With the formation of the German Jagdstaffeln, the era of colorful German fighter aircraft began. Before we present the first markings and paint schemes of the first ten Jagdstaffeln, it is important to explore the reasons for these colorful paintings.

As with my mentor Alex Imrie, my conversations with the former fighter pilots were not only about learning as much as possible about the personal painting of the aircraft, but also finding out the reasons for it. During my summer vacation with Alex at Harpenden in England, these were our main topics of conversation. Summarizing the results of our research and discussions, it was primarily two military considerations, along with personal reasons, that led to the colorful paint schemes of the fighter planes.

5.1 Identification of the Aircraft of the Pilot's Own Jagdstaffel

When operating as a Jagdstaffel with 12 aircraft in groups of six aircraft or in Ketten of three aircraft, it was important that each pilot could immediately identify his position in the formation and that of the other pilots. The Staffelführer and Kettenführer (leader of three aircraft) had to know where each pilot was at all times and had to keep track of whether all aircraft were still present or if one was missing.

Josef Jacobs, Staffelführer of Jagdstaffel 7, told me the following in relation to this:

"I regularly left my usual position at the head of my Staffel and climbed higher and circled around my Staffel to make sure that the flight formation was maintained, and to check if all the planes were still there and none had been lost. I learned this from my former Staffelführer, (Oblt.) Erich Hoenmanns, at Jasta 22."[1]

Other Staffelführers or deputy Staffelführers like Heinz Arntzen, Jagdstaffel 50, Walter Böning, Jagdstaffel 76, Hans Holthusen, Jagdstaffel 29, Werner Junck, Jagdstaffel 8, Hans Jungwirth, Jagdstaffel 78, Karl-August von Schoenebeck, Jagdstaffel 33, Kurt Student, Jagdstaffel 9, Victor Schobinger, Jagdstaffel 12 and Rudolf Stark, Jagdstaffel 35, also reported to me that they proceeded in the same way. They, too, regularly left their position at the head of the Staffel to ensure that no one was lost and that the flight formation was maintained.

When hostile aircraft were encountered, a successful outcome often depended upon the attack being made quickly and in a closed Staffel, Halbstaffel (six aircraft) or Kette (three aircraft) formation. Accordingly, the Staffel- and Kettenführer had to be sure that each pilot maintained his assigned position and that the Staffel attacked together. If this attack was carried out in a disciplined manner, the first success could be achieved by breaking up the enemy flight formation. In particular, the two-seater units, which were often flying in close formation, were extremely difficult to attack, since there was only a small blind spot that was not covered by the machine guns of the two-seaters. This was especially true if the two-seaters managed to form a defensive circle. Thus, it was in the manner of the attack to what extent the attacking fighters flew into a hail of enemy shells or were at least partially spared from it. If the first attack succeeded in breaking up the enemy formation, individual dogfights began, often spread the aircraft over a larger area.[2]

These flight maneuvers were done at altitudes ranging from a few hundred meters to 5,000 meters, in gusts of wind, clouds and rain, and in a wide variety of daylight and weather conditions. This placed demands on the Staffelführer, most of whom were only twenty to twenty-two years old, that are almost inconceivable today and that **represents** an almost unbelievable achievement!

An example of the conditions under which a mission and the following air fights took place can be seen in the report and altimeter recording of Lt. Walter Böning, Jagdstaffel 19, dated August 16, 1917. When I photographed his photo album, the record of the altimeter of his Albatros D V with the above date caught my eye. During my following visit I asked him why he had pasted this record into his photo album.

For quite some time we sat together, and he told me the story behind the altimeter record in a very lively manner:

Since 26 June 1917, Jagdstaffel 19 had been based at St. Loup airfield, south of Chateau Porcien in the

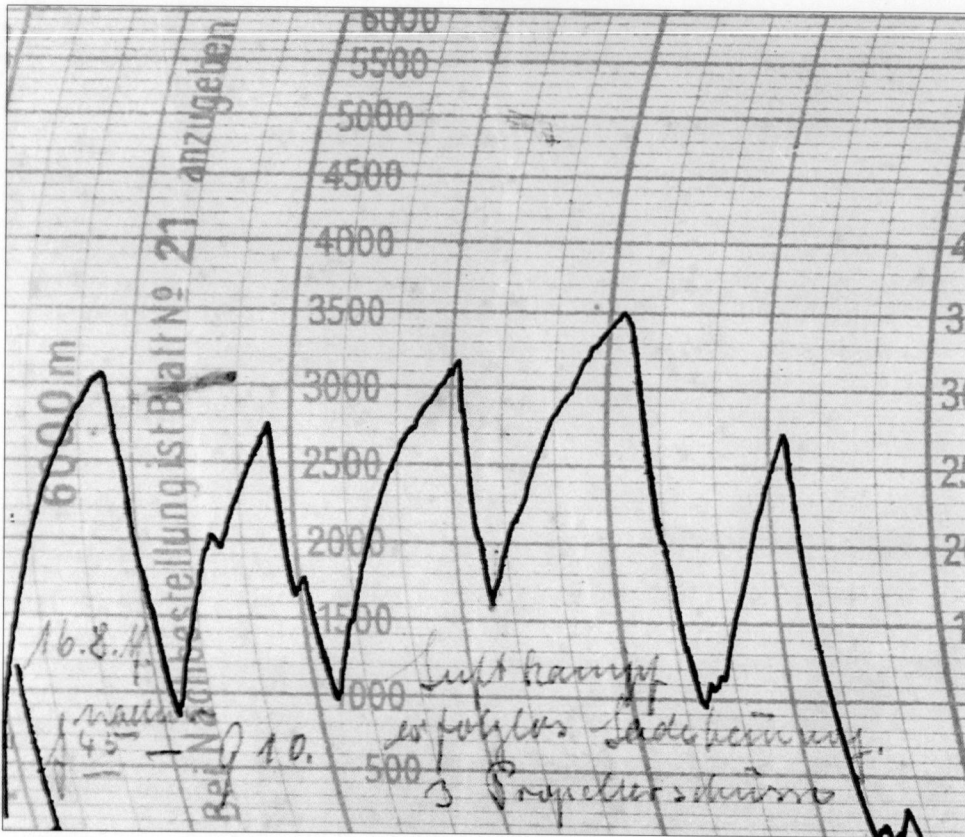

Left: The document recorded by Lt. Walter Böning's altimeter (Jagdstaffel 19) during his sortie on 16 August 1917. This illustrates how often the altitude was changed and gives a good idea of the physical and psychological demands of an air combat. The handwritten annotations inform us that the flight took place in the morning between 7:45 and 9:10, was unsuccessful due to jammed machine guns, and that his propeller was hit by three shots.

Champagne, in the area of the German 7th Army. The Staffelführer at that time was Oblt. Erich Hahn.

"On 16 August 1917, Erich Hahn, Arthur Rahn, I and two other pilots from the Staffel, took off from our airfield at 7:45 a.m. After takeoff, we turned west toward Neufchatel to gain altitude, taking the advantage of the westerly wind. Then we turned south in the direction of Reims gaining further altitude.

After about 10 minutes we had reached an altitude of 3 000 meters. Over the front east of Reims I recognized a formation of four French SPAD fighters below me. I tilted my plane down and I saw that my companions were also attacking the SPAD in a dive. I opened fire on one of the French fighters, which then dipped down. In the dive I pursued the SPAD to an altitude of about 800 meters. The Frenchman then pulled his plane up and I tracked him to 2 500 meters. At this altitude, the SPAD turned down again with the intention of shaking me off. I went into a dive again, but lost sight of the enemy plane in the ground haze at about 900 meters.

I then pulled my Albatros back up to find my comrades and continue searching for the enemy aircraft. Arthur Rahn appeared next to me. Shortly thereafter we spotted Oblt. Hahn's machine and joined him. A little later the other two Albatros also appeared. We five then continued the front flight. About ten minutes later we had reached about 3,200 meters, when we again recognized French SPAD below us. Oblt. Hahn gave the signal to attack and, followed by Arthur Rahn and myself, again we went into a dive, while the other two pilots stayed above us to cover us against a possible attack from above.

I reached the enemy aircraft at about 1 600 meters and opened fire from both machine guns. Instead of letting itself spin down, as I had expected, the SPAD climbed, pursued by me, to about 3 500 meters, and then attacked me in a turn from above. It came to a wild dogfight in which we went down to about 1 000 meters. Then the SPAD pulled up again to 2 700 meters. There he tried to attack me again in a turn but got right in front of my MG's. I tried to shoot, but both MG's jammed. This was the moment for me to break off the dogfight immediately! I immediately turned to the north and went down in a dive. The SPAD kept firing behind me until I was almost at treetop height. Suddenly the machine gun fire in my rear stopped. I looked around and the SPAD had disappeared. Instead, Hahn's Albatros appeared behind me, followed by Rahn's plane. Hahn had recognized my precarious situatio and pursued the SPAD in turn, whereupon the latter let go of me, especially since we were at

Right: Lt. Otto Kissenberth in the pilot's seat of his Albatros D II(LVG), Jagdstaffel 16, photographed by Lt. Hans Auer in February 1917.

low altitude over German territory. Flying straight northwest, I landed safely at St. Loup a short time **later,** protected by the two of them. At home I recognized that the propeller of my machine had been hit by three shots. Hahn had saved my life, for which I invited him to a bottle of champagne that evening."[3]

During the flight, which lasted around 1½ hours, Walter Böning had changed altitudes several times, by up to 2 500 meters. As a souvenir of this memorable dogfight, he pasted the printout of the altimeter into his photo album. In addition, he was impressed by the flying skills of his French opponent in the dogfight, which he acknowledged in conversation with the words:

"The man could fly damned well; I would have liked to know who he had been".[4]

A good Staffelführer like Oblt. Erich Hahn broke up their own attack in order to rush to the aid of their Staffel mate:

Josef Jacobs put it to me like this:

"This was the difference between a good Staffelführer or a reckless "Abschießer" (egoistic shooter). I experienced the difference when my Staffel was first part of Jagdgruppe 11 and then Jagdgruppe 6.

The leader of Jagdgruppe 11 (consisting of Jagdstaffeln 7, 29, 33, and 35, author's note) was Oblt. Schmidt Staffelführer of Jagdstaffel 29. He belonged to the first category. With all his courage and attacking spirit, he never took unnecessary risks for his pilots or the pilots in his Jagdgruppe. He always had an eye for the situation of his men and never hesitated to rush to their aid.

Quite different was the leader of the Jagdgruppe 6 (consisting of Jagdstaffel 7, 28, 57 and 51, author's note) Oblt. Gandert, Staffelführer of Jagdstaffel 51. He belonged to the second category. For him it was all about getting kills and nothing else. His losses were accordingly frighteningly high. I pointed out my opinion on this to him very clearly several times, but he was unteachable! To make matters worse, he was the leader of our Jagdgruppe, and, as you can imagine, our relationship was bad according to this, because he, an active Oberleutnant, hated to be criticized by me, a Leutnant of the reserve. But that was absolutely indifferent to me and finally, Thuy (Staffelführer Jagdstaffel 28) and I did what we want and ignored his orders."[5]

For this reason, the performance of a Staffelführer must not only be measured by the number of aerial victories of his Staffel, but the successes must be seen in relation to the losses of his Staffel in air combat. A very good Staffelführer achieved high victory rates while suffering only low losses in the process. An example for this is Lt. Josef Jacobs, Staffelführer of Jagdstaffel 7 or Lt. Hermann Frommherz, Staffelführer of Jagdstaffel 27, who achieved great successes with very few losses among their own pilots.

In this context, one must also consider the enormous physical and psychological strain on

Above: Offz.Stellv. Sturm takes off in his Albatros D II D.1772/16 marked with the identification "E" from the Jagdstaffel 5 airfield at Boistrancourt.

Below: Colorized photo of Albatros D II D.1772/16 taking off. The black "E" is only visible from a distance due to the white border on the honey-colored fuselage. A black "E" without a white border would have been difficult to see. (Coloring: Aaron Weaver)

Above: Lt. Kleemann, Jagdstaffel 5 waves to his comrades on the ground during takeoff with his Albatros D II D.1740/16 from Boistrancourt airfield.

Below: The colored photo shows how well the red "F" on the honey-colored fuselage was visible even from a distance. (Coloring: Aaron Weaver

the fighter pilots and especially the Staffelführer, which Josef Jacobs impressively described to me in conversation:

A dogfight is like riding a roller coaster for 20 minutes with constant turns, rolls and rollovers. After such a dogfight you are without any orientation. You don't know where up and down or right and left is. In this situation, you have to recognize the planes of your own Staffel as quickly as possible to join them. My task as Staffelführer was to gather my group as quickly as possible and to restore flying formation again.[6]

A pilot who had lost his orientation even for a very short time and therefore did not come under the protection of his Staffel was exposed to the highest danger. As former fighter pilots reported to Alex Imrie and me there were "specialists" on both sides of the front, who did not take part in the dogfight,

but waited high in the cloud until it was over. They were lying in wait for a "bunny" who had lost his orientation, and then attacked with lightning speed from the clouds and shot down the aircraft. Before the other pilots of the Staffel realized what had happened, the "specialist" had already disappeared safely in the direction of their own lines.

For these reasons, it was of the utmost importance, even a matter of life and death, after an air fight to recognize one's own aircraft as soon as possible, to join them and, if necessary, to take up a defensive formation. It was the task of the Staffel- or Kettenführer to group his aircraft as quickly as possible after an air fight. If the own planes were colorfully painted, they could be recognized much faster and easier.

Josef Jacobs related in this context:
"Already at Jagdstaffel 22, (late 1916 – early

Above: The Albatros D II (L.V.G.) D.1043/16 of Jagdstaffel 23 at the airfield of Erlon, in April 1917. The Staffel-marking is the Germanic good luck symbol "Swastika" in black.

Below: The colored photo of the Albatros D II D.1043/16 shows that in an air fight the black "Swastika" have been difficult to recognize. For this reason, this Staffel-marking was no longer used when the units received their new Albatros D III. (Coloring: Jim Miller)

Above Left: Vzfw. Heiliger's Jagdstaffel 30 in flight on his Albatros D III D.2126/16 with the personal identification "X" in March/April 1917. (G. van Wyngarden)

Above Right: The colored photo gives a good impression of how the aircraft looked in the air. The black "X" can only be seen because of the white border. Without the white border, the black "X" would be difficult to recognize. (Coloring: Aaron Weaver)

1917, author's note) we painted the planes differently. This originated with our Staffelführer, Oblt. Hoenmanns, who wanted to recognize the individual planes in the air."[7]

Josef Jacobs added:

"As a Staffel- or Kettenführer, you had to know every paint scheme of your Staffel's aircraft in your sleep. You had to recognize when someone was in danger and whether it was a 'bunny' you had to rush to help or an experienced pilot who could handle the situation himself. You had to check if someone was missing and if the flight formation was kept.[8]

The statement of Josef Jacobs was also confirmed to me in conversation and correspondence by other Staffel- and Kettenführer. The astonishingly good memory of most fighter pilots concerning the painting schemes applied to their airplanes and those of their comrades was due to the fact that it was of utmost importance to know the insignia and painting of their comrades. These were memories that were not forgotten and were very well recalled when the correct questioning technique was used.[9]

Since the aircraft in the Jagdstaffeln usually flew staggered according to altitude, some Staffelführer also placed individual marking on the top of the fuselage and on the upper wing.

When Jagdstaffel 5 received their new Halberstadt fighters in autumn 1916, the Staffelführer Oblt. Hans Berr gave the order that the aircraft had the personal marking, a number, or a letter not only on both sides but also on the upper part of the fuselage. This enabled the pilot flying behind to recognize which aircraft was in front of him due to his higher position.[10]

During the course of the war, it became apparent that the elevation control surfaces were an excellent place for the Staffel marking. As a result, the vast majority of German Jagdstaffeln placed their unit-markings on the elevation control surfaces.

Unfortunately, with the exception of a few French photos of captured German airplanes exhibited in Paris around the end of the war, almost all photos of WW I aircraft were taken in black and white. For this reason, I requested Aaron Weaver and Jim Miller to colorize some in-flight photos in order to provide an impression how these planes were perceived in the air by friend and foe alike.

Above: Lt. Strey of Kest 1b in the pilot's seat of his Albatros D III (O.A.W.) in flight.

Below: The colored photo shows how well the blue fuselage and the blue tail with the white edge can be seen in the air. Also the personal identification a white "S" is very clearly visible. (Coloring: Aaron Weaver)

Above: Lt. Wewer of Jagdstaffel 26 with his Albatros D III on approach to the Bohain airfield in the area of the German 2nd Army.

Below: The colored photo illustrates how well the black and white striped fuselage was visible in flight. (Coloring: Aaron Weaver)

Above: Lt. Paul Strähle, Staffelführer of Jagdstaffel 57 in flight with his Albatros D III (O.A.W.) over an industrial area. The photo was taken by his comrade Lt. Johann Jensen.

Below: The Staffel-marking, the light blue fuselage, and the personal marking, the red airplane nose, are very well recognizable from the aircraft flying along. (Coloring: Aaron Weaver)

Above: An Albatros D III of Jagdstaffel 27 in flight.

Below: The colored version of the Albatros D III is interpreted as black and white.

5.2 Confirmation of Aerial Victories

As already explained the task of the Jagdstaffeln was to protect their own army cooperation aircraft to fulfill their duty unharmed and to weaken the enemy air forces by shooting them down or forcing them to land. Accordingly, the performance of a Jagdstaffel was measured primarily by the number of aircraft it shot down or forced to land on its own territory. To evaluate the performance of a Jagdstaffel, it was necessary to record the enemy planes that were brought down and to attribute these victories to the Jagdstaffel and the airmen who achieved the victory.

The successes in air combats were published by all countries involved in the war. One reason for this publication was to bolster the morale of the own troops and to motivate the fighter pilots to take higher risks in combat. The higher the number of aerial victories, the higher was the probability for a fighter pilot of being mentioned in the army report, of being promoted to a higher rank and receiving recognition by being awarded much sought-after decorations.

5.2.1 The German Fliegertruppe

There were clearly defined rules for the recognition of an aerial victory in the German air force:

1. A downed enemy aircraft was awarded to only one airman or crew.[1]
2. The reported aerial victory had to be validated by the wreckage of the crashed plane or the intact landed enemy aircraft.
3. For the recognition of the aeriel victory to a certain Jagdstaffel or a certain pilot or crew, witnesses had to be named. These were usually the Luftschutzoffiziere (air defense officers), members of the infantry, artillery, anti-aircraft units or balloon observers. High importance was placed on confirmation provided by witnesses from the ground. The testimony of the fighter pilot who claimed the kill and the testimony of another fighter pilot who had observed the kill were not sufficient to be awarded confirmation of an aerial victory.
4. If the enemy aircraft crashed over no-man's land or behind enemy lines, the shootdown had to be confirmed by eyewitnesses in the frontline area.
5. If it was not clear from the statements of the ground troops or crew of the captive balloons whether the enemy aircraft had been destroyed, the aerial victory was recognized as: "forced to land". However, this recognition was not considered a full-fledged aerial victory.
6. If several pilots or crews of different units claimed the same kill, an arbitration panel was formed to decide on the allocation of the aerial victory. The practice of "shared victories", that was common in the flying services of other nations, was not taken up within the Fliegertruppe.
7. If more than one pilot of a unit claimed the same shot down hostile aircraft, the victor was usually decided by draw.

When a German pilot returned from a frontline flight and filed a report of shooting down a hostile aircraft, it was the task of the Officer z. b. V. to track down appropriate witnesses. For this he used

Exemplary illustration of the telephone network of the Jagdstaffeln in an army section of the German 2nd Army.

Right: Enemy aircraft forced to land always generated a great deal of attention for the ground troops, especially if they came down in one piece in an area safe from shelling. On 17 May 1916, German troops salvaged F.E. 2b 6341 from 25. Sqn. RFC, which was brought down on the previous day near Fournes. The crew consisting of Capt. D. Grinell-Milne and Cpl. D. McMaster were captured by German forces.

Einbringen e. feindl. Flugzeuges: 17.5.16

the special telephone network available to the Jagdstaffel, about which the " Weisungen für den Einsatz und die Verwendung von Fliegerverbänden innerhalb einer Armee" has to say:

"(No. 47.) The airmen require a special telephone network in order to be kept constantly informed of enemy air activity....

(No. 48.) The leadership of the air defense service is the responsibility of the KoFlak (commander of anti-aircraft-artillery) and Flakgruppen-Kommando (Commander of anti-aircraft-formations). The Jagdstaffeln, the "Gruppenführer der Flieger" and the "Kommandeur der Flieger" are to be connected to the air defense network by direct lines.

(No. 49) In order to ensure the cooperation of flak and airmen, it is advisable on battle fronts to assign an Luftschutzoffizier (air defense officer) to the flak center and to set up a special observation post with a good view in the vicinity of the flak center. The Luftschutzoffizier, equipped with binoculars and rangefinder, is to receive all reports of enemy air activity from the troops. He will check the reports for accuracy, obtain a picture of the situation in the air from them and from his own observations, and keep the Staffelführer of the Jagdstaffeln informed of their own and enemy aviation activity in their section.[2]

In the instruction for the deployment of Jagdstaffeln, the telephone connections of the Jagdstaffeln were shown as an example. The Jagdstaffeln were directly connected with the Gruppenführer der Flieger and through him with the Luftschutzoffizier. The later formed Jagdgeschwader were in direct telephone contact with the Army High Command (A.O.K.).

The first contact of the Offizier z. b. V. in seeking confirmation of the aerial victory was usually the Luftschutzoffizier of the area where the aerial combat had taken place. If the Luftschutzoffizier was unable to personally confirm the claim, he was able to forward the Offizier z. b. V. to the anti-aircraft units, infantry and artillery, or the balloon platoons located in the area of the presumably shot down hostile plane. If witnesses were found who confirmed the claim, they wrote a report about their observations. These reports, together with the reports of the pilot and other witnesses, were submitted to

the Kommandierenden General der Luftstreitkräfte for examination. From there, the confirmation of the aerial victory was granted – or not.

The rules for the confirmation of an aerial victory were kept very strict. A number of fighter pilots I spoke to had to experience that their aerial victories were not confirmed due to a lack of eye-witness reports, or were awarded to other pilots or the anti-aircraft artillery.

Fritz Jacobsen, former member of Jagdstaffeln 17, 31 and 73, whom I visited several times in Nuremberg, also experienced this. He reported 11 aerial victories of which only 5 were confirmed, and this was by no means an isolated case.

With Jagdstaffel 31 he had two confirmed aerial victories: One was a B.E.2 brought down inside German lines on 6 July 1917 and the other was a Sopwith single seater on 19 August 1917 which was brought down inside German lines. Two further claims mentioned in the war diary were not confirmed: These were on the Italian front, a Nieuport near Santa Lucia on 28 September 1917, and an S.A.M.L. near Feistritz on 26 October 1917.[3]

As a member of Jagdstaffel 73 he filed claims for six aerial victories, of which only two were eventually confirmed. I visited Fritz Jacobsen several times and he was absolutely no showoff.[4]

As long as only one or two Jagdstaffeln were stationed in an Army area, the small number of available aircraft meant that shooting down an enemy aircraft was not such a frequent occurrence, unless it occurred in a major combat area. The confirmation of an aerial combat by witnesses from the ground was therefore rather easy to obtain. This was one of the reasons why some of the Jagdstaffeln initially did not consider it necessary to apply identification markings that were clearly visible from the ground to their aircraft.

With the permanent increase in the number of aviation units at the front, the number of aerial encounters also increased steadily. By the spring of 1917, the air fights in certain sections of the front increased considerably and there were battles in the sky involving several Jagdstaffeln on both sides.

According to available photos, Jagdstaffel 29, then under the command of Oblt. Fritz Dornheim, was in February/March 1917 one of the first if not the first Jagdstaffel which used special markings in order to be well-recognized from the ground. For this reason, the numbers and letters the aircraft where not only applied to both sides and onto the decking of the fuselage, but also onto the fuselage bottom and bottom of the lower wings.

By 1918 it was not uncommon that a hundred or more aircraft were involved in an aerial encounter, which made it very difficult to attribute an aerial victory to a particular airplane and pilot. It became apparent that the more striking a single aircraft or the aircraft of a Jagdstaffel were painted, the greater the chance was that the aircraft would be recognized from the ground after a successful air fight.

Josef Mai, a former member of Jagdstaffel 5, who knew of my interest in the painting of fighter planes, explained to me:

"The planes of our Staffel became more and more colorful and this gave us the advantage when we needed confirmation of an aerial victory. The soldiers on the ground could see our colorful birds very well. Especially after we painted large letters on the lower wing (bottom of the lower wing, author's note).[5]

Josef Jacobs, Jagdstaffel 7 told me about this topic:

"Our black birds were quite well known by our troops in Flanders, especially after we had also painted the underside of the wings black and a large white letter was painted on the underside (Usually the first letter of the airman's family name, author's note). *After that it was easier to obtain confirmation, because the big letters were easily recognizeable.*

We also regularly descended to 200 to 300 meters and saluted our troops on the ground by wobbling our wings as we flew over. Soon we were known as the "black bunch" and at some point, even came up with the name "Geyers schwarzer Haufen."[6]

"Geyers black bunch" was a group of feared outlaws from the time of the German peasant wars in the 16th century who fought against the church and the nobility.

5.2.2 The British Military Air Service

On the British side, no special achievement was initially seen in the shooting down of an enemy aircraft. It was not until the achievements of successful fighter pilots were brought to public attention in France and Germany that the British Royal Flying Corps (RFC) and the Royal Naval Air Service (RNAS) also began to register aerial victories and present successful pilots to the public.

When registering an aerial victory, the British Air Services primarily relied on the testimony of the pilot or crew. Their testimony about the shooting down of a German aircraft was documented after the return of the aircrew and was then submitted to the "Intelligence Officer". When the report was accepted by the latter, it was published. The RFC, the RNAS and later the Royal Air Force (RAF), did not have an organization that checked the validity of aerial victories, as was common on the German or French sides.

A special feature of the "British system" was that, in addition to the report "destroyed" of an enemy aircraft, there was also the "out of control" category for a claim. These reports were based on the British pilot's assumption that the German aircraft went down in an uncontrolled manner. As former German fighter pilots explained to me in unison, it was part of the standard repertoire of the fighter pilot to let himself spin down over his own territory in the event of an unfavorable position in an air fight. This was a rather successful method of extracting oneself out of a critical situation since allied planes usually broke off their pursuit at about 1,500 meters in order to avoid coming under fire from German anti-aircraft artillery or machine guns.

Accordingly, recorded German losses that match the majority of these "out of control" claims are hard to find. This is also due to the fact that all sides only registered "bloody" losses, i. e. those with injured or killed crews. A German pilot or crew that was forced to make an emergency landing could be re-deployed if they were not injured; an aircraft damaged in the emergency landing could be repaired and return to frontline service soon enough.

5.2.3 The French "Aéronautique Militaire"

The French Aéronautique Militaire employed rules for confirmation of aerial victories that were similarly strict to those of the German side. An aerial victory had to be confirmed by three witnesses who did not belong to his unit, and could be pilots

Above: Lt. Höhne in the cockpit of Airco DH 2 7873 forced to land by Hptm. Oswald Boelcke on 14 September 1916. Standing on the wheel of the aircraft and talking to him is Hptm. Boelcke himself. Again, a number of curious admirers can be seen in the photo. German pilots were eager to fly captured enemy aircraft if they were brought down in airworthy condition in order to learn about the flight characteristics of these aircraft, and to analyze the weaknesses of each type.

of another unit, ground troops or balloon observers. The reports were verified, as on the German side, and an aerial victory was acknowledged or not. If it was not certain that the German aircraft had actually been destroyed, the claim was registered as "probable" (possible), but like the "aircraft forced to land" on the German side, it was not considered a full air victory. One difference to the German system was that the aerial victory could be awarded to several French pilots or crews involved in the shooting down of the German aircraft, which meant that each was credited with an aerial victory.[7]

5.3 The Different Air Combat Conditions

In this context, one must also consider the conditions under which the air fights took place in the various frontline areas.

British Front

The leadership of the British RFC and RNAS pursued an "offensive strategy". This ordered the aircrews always and under all circumstances to fly over the lines in order to attack the German planes over their own territory.

For this reason, a large number of the air fights on the British front took place in the frontline area or over German territory. As a consequence, a high percentage of British aircraft crashed or came down in emergency landings over German territory, or were shot down within sight of German troops, which greatly favored confirmation of German aerial victories.

As a result, German air victories in the British sector are easier to trace. An example is Jagdstaffel 30, which was deployed during the entire time of its existence in the area of the German 6th Army and was involved in combat with British aircraft exclusively. Of the total of 64 confirmed aerial victories, 57 are clearly attributable to British losses. This amounts to 89% of the aerial victories confirmed to the Staffel.

Above: Allied aircraft captured intact, such as this Nieuport 17 donated to the Fliegertruppe by French Escadrille Spa 48, were marked with German national insignia for evaluation by the Jagdstaffeln and at Armee-Flugparks and Jagdstaffelschulen.

In contrast, British crews faced the problem that, in a significant proportion of reported aerial victories, they could not determine whether a German aircraft they were engaged in combat with, actually crashed or just ended the encounter by spinning down. The British pilot, after all, could not land on German territory and ask for confirmation of his victory. For this reason, the report he subsequently filed concerning shooting down a German aircraft depended largely on the perception and subsequent account of the pilot. Accordingly, the problem arose that by summing up several individual reports, one and the same German loss was registered and confirmed several times, or even that a British aircraft that was going down was mistaken for a German aircraft crashing within its own lines.

French Front

The French air forces pursued a more defensive strategy, which was primarily concerned with securing their own airspace. For this reason, a large part of their air combats took place over the front or over French territory.

In this case the German Jagdstaffeln found themselves in a position comparable to that of the British crews on the German front. They, too, could not land and inquire whether a French aircraft they had taken under fire had actually crashed or whether the French pilot had escaped in a spin. This made obtaining confirmation of German aerial victories on the French front much more difficult.

5.2.6 Summary

Comparing reported aerial victories and losses, it can be seen that, for all the inevitable errors and inaccuracies in confirming aerial victories, the rigorous German, and French systems, with their systematic verification of the combat reports, proved by and large to be the most reliable.

After more than 40 years of studying of hundreds of German, British and French combat reports as well as many interviews with former fighter pilots, I am convinced that the **vast majority of the fighter pilots of all nations** honestly reported what they saw or believed they saw under the stress of an air combat. Of course, there was the odd dubious individual in each flying service who tried to obtain information for aerial victories that had never happened. However, the British system made the career of a Canadian fighter pilot possible whose claims for aerial victories seem highly questionable when matched against German losses.

Considering how differently air victories were registered and confirmed or not confirmed by the various warring nations, as well as the different conditions under which the air battles took place over the various front lines, the repeatedly published rankings of the "aces" according to the number of their aerial victories must be viewed with a certain degree of skepticism. Besides this, a war, even when fought in the air, is not a sporting competition, but is deadly serious.

Above: The spoils of war. The wreckage of shot down Allied aircraft piled up at a German Armee-Flugpark.

Above: Hard landing of Lt. Hans-Hermann von Budde, Jagdstaffel 29. The Albatros D III had the personal marking a black "B" with a white border on both sides, the top and the bottom of the fuselage, to be recognized by higher flying comrades as well as from the ground.

Right: The photo shows that the black "B" with the white border was also on the underside of the lower wing to be recognized from the ground troops.

Above: Lt. Strey of Ke.St. 1b on approach to Karlsruhe airfield.

Below: On the light blue underside of the wings, the marking "S" is clearly recognizable.

Above: Photo from the ground of the Albatros D Va (O.A.W.) flown by Lt. Hans-Joachim von Hippel, Jagdstaffel 5 in late autumn 1917.

Below: The colored photo shows how good the marking "H" could be seen from the ground. (A. Imrie)

Above: Lt. Ernst Udet, squadron leader of Jagdstaffel 37 flies his Albatros D Va in a loop over the airfield of Metz. (A. Imrie)

Below: The large white "U." can be clearly seen on the left wing underside. The two pennants on the tailplane are his identification as Staffelführer. (A. Imrie)

Above: Two Albatros D II (at left) and five Albatros D III of Kest 1b at Mannheim airfield. All aircraft have a white letter identification on the upper wing. The same identification was also on the underside of the lower wing in black.

Right: Two Fokker D VIIs of Jagdstaffel 7 with the markings on the upper wing described by Josef Jacobs. The same marking was on the lower side of the lower wing to be visible by the troops on the ground.

6. The Letter- and Number-Markings of the First German Jagdstaffeln

Analyzing the available photographs taken in 1916 of the first ten German Jagdstaffeln, one comes to the conclusion that the identification system previously used by various Kampfstaffeln served as the role model for their method of choosing markings for the new single-seat fighters. This may also have been due to the fact that some of the Staffelführer and pilots had previously served with these Kampfstaffeln, so "importing" a proven system to the new units clearly made sense.

Jagdstaffeln 6 and 10 used numbers as means of identification for their aircraft, while Jagdstaffel 5 employed a combination of numbers and letters.

In the case of Jagdstaffeln 1, 2, and 4, letters were in use. However, the letter identification was not a fixed system and not used by all pilots. The pilots of these Staffeln enjoyed considerable freedom concerning the choice concerning the personal marking that was to be applied to their aircraft. If they used letters, they were not arranged alphabetically but usually symbolized the first letter or letters of the family name.

Almost all photos taken during World War I were taken on "black and white" filmstock.[1] This resulted in the letters and numbers of the aforementioned aircraft almost uniformly being interpreted as black in previous publications. One reason is perhaps that it is tempting to summarize that markings that look black in these photos must simply have been black. However, based on my long-time research I cannot agree with this interpretation for the following reasons:

1. According to the statements of former airmen I interviewed whose Jagdstaffeln used letters or numbers for identification of their airplanes, black numbers or letters were used only to a very limited extent. The black color was usually chosen only when applied to a white or very light-colored background (e.g., sky blue fabric covering or surfaces painted white, light blue, or a light grayish white).[2] Alex Imrie confirmed these statements to me based on his interviews with former airmen.

For my first visit to Herr Rudolf Nebel, Jagdstaffel 5, I had painted side sketches of the Halberstadt fighter planes with colored pencils. I had drawn the number that had been applied to the Halberstadt with the light fabric covering in black, simply following the examples shown in publications

Above: Rudolf Windisch poses in front of an O.A.W.-built Albatros D.II marked with the number "2". This picture illustrates the difficulties of interpreting the details of an aircraft from a single available photo. The fuselage can be interpreted as being light-colored varnished or stained reddish-brown plywood, both seem possible. As a consequence, the number "2" could be either yellow, red or black, all of which render very dark on orthochromatic film. It is not known if the airplane was the personal mount of Windisch, and, unfortunately, the military number of the plane marked on the wheel cover cannot be deciphered.

available at this time. But Herr Rudolf Nebel corrected me, stating that the numbers had been applied in red. The reason was that: *"Otherwise they would not have been seen well in the air"*.

2. The intention for the application of the numbers or letters was that a fast-moving aircraft in the air could then be identified by other pilots even from a distance. A white or yellow number or letter is easily

Albatros D II (O.A.W.) with a black number, Profile 19

Albatros D II (O.A.W.) with a yellow number, Profile 19a

recognizable on a dark background such as a rust-red wooden fuselage or dark blue-gray canvas. A black number or letter, on the other hand was not.

Even on a light background, a red number or letter is considerably easier to see than a black number or letter. In addition, one must consider that next to the number or the letter was a white-bordered, deep-black Iron Cross. Next to this cross a black letter or number would simply disappear from a distance.

3. In this context it is illogical that a Staffelführer would order an identification system to be applied to the aircraft of his unit and then use a color that is very difficult or almost impossible to recognize while in the air!

4. Finally, by looking at the photos, it must be taken into account that the color red and certain tones of yellow appear black on photos taken with the orthochromatic film used at that time. It is then not

possible to distinguish the color black from red or yellow.[3]

Visibility of Different Colored Letters on Different Aircraft Fuselages

For the following color profiles, we used a photo of an Albatros D II (O.A.W.) marked with the number "2", with Lt. Rudolf Windisch posing besides the aircraft.

This photograph has been chosen to illustrate the difficulties in interpreting a single photo when no additional information concerning the subject is available. To begin with, O.A.W., then still called "Albatros Werke Schneidemühl (A.W.S.), only manufactured a batch of 50 Albatros D II fighters. Details of these are not exactly perfectly documented in photos and literature. To add to the confusion, some were delivered with "plywood" fuselages, such as our example shown here, while

Albatros D II (O.A.W.) with a black number, Profile 20

Albatros D II (O.A.W.) with red number, Profile 20a

others were delivered with the fuselage carrying a two- or three-toned camouflage scheme. An example illustrating the latter scheme is also featured in this book in the form of Albatros D II (O.A.W.) "9", flown by Jasta 6 pilot Carl Holler.

In the case of the plywood-fuselage D IIs manufactured by O.A.W., it is not entirely clear if these were given a clear, transparent coat of varnish that resulted in a "light" colored fuselage or if the varnish was tinted. This would result in a "rust red" or "reddish brown" fuselage color which has been referred to by several veteran pilots. The shade of the plywood shown in the accompanying photo leaves both interpretations within the realms of possibility. Consequently, this aircraft serves as an example to illustrate how the visibility of differently colored numbers varies on backgrounds of different colors, in this case the varnished plywood of two shades.

All known details that were specific to the O.A.W.-built D.II have been faithfully incorporated into these profiles.

Albatros D II (O.A.W.) with a Black Number, Profile 19

The color profile illustrates the visibility of a black number on a rust-red fuselage of an **Albatros D II (O.A.W.)**.

Albatros D II (O.A.W.) with a Yellow Number, Profile 19a

The color profile illustrates the visibility of a yellow number on a rust-red fuselage of an **Albatros D II (O.A.W.)**.

Albatross D II (O.A.W.) with a Black Number, Profile 20

The color profile illustrates the visibility of a black number on a honey-yellow fuselage of an **Albatros D II (O.A.W.)**.

Albatross D II (O.A.W.) with Red Number, Profile 20a

The color profile illustrates the visibility of a red number on a honey-yellow fuselage of an Albatros D II (O.A.W.).

Taking into account the statements of the former members of the Jagdstaffeln, the following summary can be made concerning the color of the numbers and letters used as identification, although this compilation does not claim to be exclusive or complete:

- **White numbers and letters** were used on aircraft with dark fuselages e.g., the Albatros D II and D III with a rust red varnish, Halberstadt D II with a dark grey-blue fabric covering and the rust red/dark green camouflaged Halberstadt D V (Han.).
- **Yellow numbers and letters** were used on aircraft with dark fuselages e.g., Halberstadt fighters with "rat-brown" or perhaps a gray-blue covering and the Albatros D II and D II with a reddish-brownish varnish.

- **Red numbers and letters** were used on aircraft with light colored fuselages, such as the Albatros DI, D II and D III with the "honey yellow" fuselages.
- **Light blue numbers and letters** were used on aircraft with dark fuselages or camouflage paint schemes, such as the Pfalz D XII operated by Jagdstaffel 32.
- **Dark blue or green numbers or letter** could be used on aircraft with light-colored fuselages, like aircraft with light yellowish linen-covered fuselages, or the "honey yellow" varnished fuselages of the Albatros D I, II and D III.
- **Black numbers and letters with white border or on white background** were used on dark as well as on light fuselages.
- **Black numbers or letters without a white background** were used very rarely and then only on very light backgrounds (white, eggshell, light sky blue).

7. The Markings and Painting of the First 10 German Jagdstaffeln, Autumn 1916 – Winter 1916/1917

At no time during World War I were there any regulations issued by the German head of the Fliegertruppe in regard to the Staffel markings of the aircraft of the Jagdstaffeln or any other aviation unit. In most cases, it was the Staffelführer who determined the Staffel marking. Most of the pilots were then free to choose whichever individual marking they then applied to their aircraft in addition to the unit marking.

Usually, there was also a "gentlemen's agreement" which regulated who used which identification or color on the aircraft. This was called "first come, first paint", which meant that the pilot who had the marking or color on his aircraft first had the right to keep it, regardless of his military rank.

Otto Fuchs was a member of the artillery and accordingly he wanted to paint the fuselage of his Jagdstaffel 30 Albatros D III black when he joined the unit in June 1917. The reason was that the artillery had black cap bands. However, Jasta 30 member Lt. Joachim von Bertrab, also an **artilleryman**, had already decided to paint the fuselage of his Albatros D III in black. As a consequence, Otto Fuchs reconsidered and opted to use green as his personal color. This symbolized his homeland, the "green Palatinate", so called because of its large forests.

Since Lt. Paul Erbguth as a "royal Saxon" claimed the Saxon colors green and white for himself, both agreed that Paul Erbguth would used a light green in combination with white and Otto Fuchs an "emerald-green" for the painting of his airplane.[1] This example shows that beside different colors, different hues of the colors were also used to distinguish the aircraft.

The following presentation of the markings of the aircraft of the first ten Jagdstaffeln does not claim to be complete. It present examples of the markings and paintings for these Jagdstaffeln at this time.

What follows is an exemplary overview of the identification markings used on the aircraft of these Staffeln from the late summer 1916 to the winter of 1916/1917 that is only based on the information of former Staffel-members, photos and documents available at this time. Sources for the presentation of the aircraft markings are presented at the beginning of the chapter.

Previously unknown photos of individually painted aircraft of these Jagdstaffeln taken during this period may surface at any time. Hopefully, additional photographs concerning the period will surface following the publication of this book. These would only broaden our perspective of the early markings applied to the first Jagdstaffeln.

7.1 Royal Prussian Jagdstaffel 1

Staffelführer:

Hptm. Martin Zander: 22 August 1916 – 10 November 1916

Oblt. Hans Kummetz: 18 November 1916 – 20 September 1917[1]

The most important and authentic information for the presentation of the markings and paintings of the airplanes in summer–winter 1916 originates with Franz Ray, former member of Jagdstaffel 1 and 28, as well as Staffelführer of Jagdstaffel 49. He was an early member of Jagdstaffel 1, serving with the unit from 1 October 1916 to 17 December 1916.[2]

After the Second World War he lived in Berlin, and Alex Imrie, who also resided in Berlin, took the advantage to visit and interview him several times. As Franz Ray told Alex Imrie, he had buried his photo album, medals, and documents from World War I in a metal box in the garden shortly before the Soviet troops invaded in 1945, thus saving them from possible destruction. This allowed Alex to photograph the photos and copy his records. As always, Alex generously shared his notes and information with me.

Available information on Jagdstaffel 1 is also based on the photo album of Lt. Hans von Keudell, who was transferred to Jagdstaffel 1 on 22 August 1916, and whose photos were generously made available to me by Greg VanWyngarden.

Another important source of information is the

Above: The Halberstadt D III flown by Lt. Hans von Keudell, summer 1916, marked with the letter "K" for the purpose of personal identification. (L. Bronnenkant)

extensive photo album of Lt. Raimund Armbrecht, which I was able to copy some years ago. He was transferred to Jagdstaffel 1 on 10 February 1917 and was with the Staffel until 23 August 1918, from 22 January 1918 on he was the Staffelführer.

In 1975, by referring to the members list of the "Alten Adler", I was able to contact Dr. jur. Ferdinand Zilcher, a former member of Jagdstaffel 1, who lived in Nuremberg. After a telephone conversation I met him for the first time at the end of 1976. To my disappointment he had to inform me that all his records and photos from World War I had become victims of the bombing raids in the Second World War, and that he did not own a single photo from that time anymore. Regardless, he told me a lot about his time with Jagdstaffel 1 and later with the Bavarian Fliegerabteilung 304 in Palestine. Even though he arrived at Jagdstaffel 1 on 20 April 1917, he was able provide me with some information about the paint schemes of the Albatros D II the Staffel received in late 1916, as there were still some old Albatros D IIs in service at that time. He recounted that, to his disappointment, he was obliged to fly his first front-line missions in an old Albatros D II, but soon he got a new Albatros D III.

Based on the analysis of available photographs, U.S. aviation historian Ed Ferko concluded in the 1970s that the Staffel was initially equipped with Halberstadt D fighters, which were operated alongside a number of Fokker D Is, and at least one Fokker D IV that was on strength of the Staffel as well.[3]

In August 1916, the Staffel apparently received the

Leutnant von Keudell
einer unserer erfolgreichen Kampfflieger

Above: Sanke card of Lt. Hans von Keudell. He was one of the most successful fighter pilots of Jagdstaffel 1 in 1916.

first Albatros D I to reach the front, D I D.385/16. Available photos document that Albatros D I

Lt. Hans von Keudell, Halberstadt D III, Autumn 1916, Profiles 21 (red "K") and 21a (black "K")

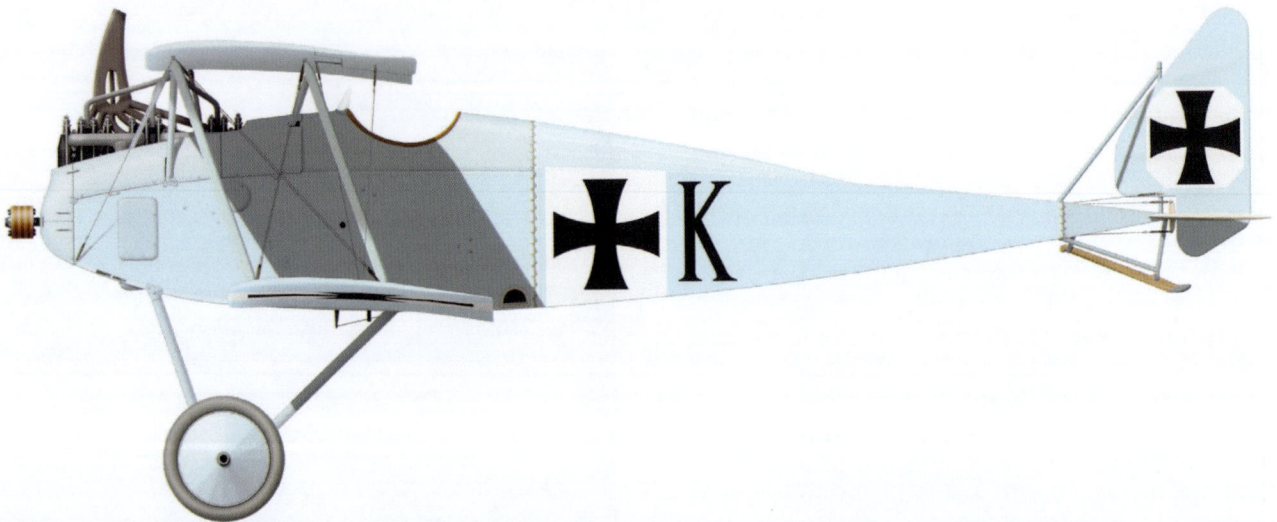

Above: The Fokker D IV of Lt. Hans von Keudell with the marking "K" on the side of the fuselage, photographed in October/November 1916. This was a very early-production Fokker D IV that may have been shipped directly to von Keudell for his personal use.

D.435/16 was also flown by the unit, probably in September 1916.

At the end of October 1916/beginning of November 1916, the Staffel was equipped with Albatros D IIs with the military numbers of the production series 1700/16–1799/16. Some of these aircraft were still in the inventory of the Staffel in April 1917.

The photos in the photo album of Raimund Armbrecht show that mainly letters were applied to the fuselages of these Albatros D II as personal markings. However, other markings, for example fuselage bands, were also used.[4]

This was confirmed by Ferdinand Zilcher. He told me that all aircraft carried personal markings when he arrived in April 1917. These consisted of letters, different colored bands, symbols, as well as differently painted fuselages. According to his recollection, there was no Staffel marking in use at that time.

Lt. Hans von Keudell, Halberstadt D III, Autumn 1916, Profiles 21 (red "K") and 21a (black "K")

His Halberstadt D III carried a "K" as personal insignia on the sky-blue fuselage fabric covering. The general interpretation is that the letter was black, which is a reasonable assumption due to the light blue background. However, the "K" may also have been applied in red, as were the personal markings on the sky-blue Halberstadt D IIIs of Jagdstaffel 5.[5] We present both the "black" and "red" versions here for comparison, so that the reader can make up his own mind which version would have been more recognizable in the air.

Hans von Keudell was born in Berlin on 5 April 1892. He entered the Bensberg Cadet School at the age of 12 and was commissioned as an ensign in the Uhlan Regiment "Emperor Alexander II of Russia" No. 3 in 1911. Already a Leutnant at the outbreak of war, he went into the field with this regiment and fought with it in France and Poland until April 1915. In May 1915 he enlisted in the Fliegertruppe and received his training as a pilot from 7 June 1915, at Flieger-Ersatz-Abteilung 2 in Adlershof. On 13 December, he was transferred to the "Brieftauben-Abteilung Ostende" (B.A.O.) in Flanders. This unit was renamed Kagohl 1 only days later and transferred to the front before Verdun at the beginning of 1916. In the summer of 1916, he was trained as a fighter pilot, and on 4 August he joined the "Kampfeinsitzer-Kommando Nord", which was attached to Feldflieger-Abteilung 32. In the course

of the formation of the first Jagdstaffeln at the end of the month, he and most of the other pilots of the Kommando were transferred to Jagdstaffel 1 on 22 August 1916. Nine days later he achieved his 1st aerial victory, to which he added nine more by the end of November 1916. For this number of victories, he would have received the Order Pour le Mérite one month earlier. However, the award regulations for the bestowal of the "Blue Max" had changed in the meantime, and now his total score of ten victories was "only" sufficient for the award of the Knight's Cross of the Royal House Order of Hohenzollern with Swords, with which he was decorated on 10 January 1917. After his 11th aerial victory, on 5 February 1917, he was appointed as Staffelführer of the newly established Jagdstaffel 27.

But his leadership was destined to be brief. On 15 February 1917, he took off with Albatros D III 2017/16. At about 17:40, after an aerial combat with Nieuport two-seaters during which he shot down one of the hostile aircraft (this victory remained unconfirmed), he – seriously wounded – made an emergency landing inside British territory near Boesinghe, where he was taken prisoner. Two days later he succumbed to his severe injuries in the military hospital at Vlamertinghe.[6]

Lt. Johannes (Hanns) Braun, Fokker D I 176/16, Autumn 1916, Profile 22

The Fokker D I was covered overall in a very light-colored linen fabric, that may have appeared almost white. As the pilot's personal insignia, the letters "HB" were applied above the fuselage military number. Braun was a native of the city of Munich, and the style of application of the "HB" monogram indicates that it was most likely inspired by the famous "Hofbräuhaus" Brewery, which is located in the heart of Munich. As would be fitting for a native Bavarian, the "HB" was most likely applied in blue color. It may also be noted that, at the time, the "Hofbräuhaus" served beer in light grey mugs onto which the "HB" logo had been applied in dark blue.

The fate of Hanns Braun is one of those stories that are never mentioned in any war chronicle, but perhaps that is why it is worth telling:

Hanns Braun was born in Munich on 26 October 1886, the son of the famous battle painter Louis Braun, whose paintings can still be admired in various Bavarian museums. He had inherited his father's artistic talent and after his school years he studied sculpture at the Munich Art Academy. Due to his enthusiasm for sports, he joined the sports club "Münchner SC" in 1902 and also played as

Left: Lt. Hanns Braun in front of his Fokker D I 176/16 with the "HB" marking, inspired by the logo of the Munich Hofbräuhaus. (R. Kastner)

Left: Ready for take-off, Lt. Hanns Braun is seated in the cockpit of his Fokker D I 176/16. (G. VanWyngarden)

a right offensive player in the football team of FC Bayern Munich. He was also an enthusiastic and successful track and field athlete and participated in the 1908 London Olympics. There he won a bronze medal in the 800-meter race and a silver medal as member of the team race. At the 1912 Olympics in Stockholm, he was again successful and won the silver medal in the 400-meter race. He set 16 German records during his active time as an athlete and was a three-time German champion in the

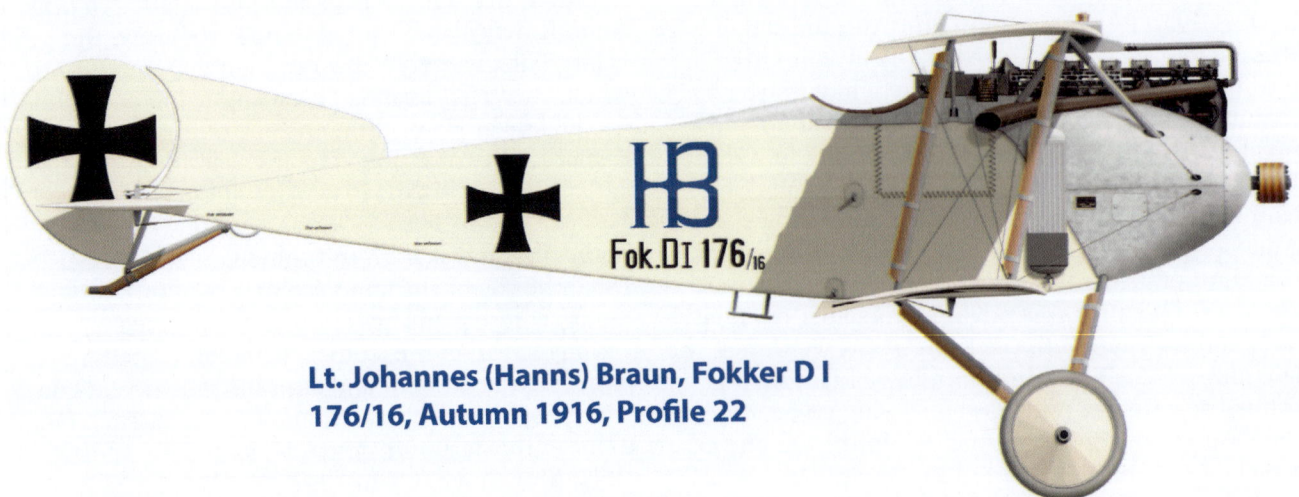

Lt. Johannes (Hanns) Braun, Fokker D I 176/16, Autumn 1916, Profile 22

Above: Photo of an original Hofbräuhaus Beer mug from around 1900. The Brewery logo likely served as the pattern for Braun´s personal marking on his Fokker D I. It is highly likely that the monogram was applied to the fuselage fabric in an identical dark blue color.

Right: Lt. Hanns Braun, in the autumn of 1918 as member of Royal Bavarian Jagdstaffel 34.

400-meter race and three-time English champion in the 800-yard race.[7] After the Stockholm Olympics, he married, and devoted himself to sculpture where he very quickly became a respected artist.

At the beginning of the war, he volunteered for military service and was first assigned to the replacement battalion of the Bavarian Infanterie-Leib-Regiment. On 20 October 1914, he was assigned to the Fliegertruppe and ground crew of the Bavarian Feldflieger-Abteilung 1. On 15 December 1914, he was promoted to Unteroffizier.

Due to an illness, Braun had to be transferred first to St. Quentin on 8 March 1915, and later to the military hospital at Bad Ems. On 9 April 1915, he was admitted to the convalescent home at Dachau for final recovery.

Since Braun had applied for training as a pilot, he was transferred to the Flieger-Ersatz-Abteilung at Schleißheim after his recovery on 12 June 1915. From 19 June to 6 August 1915, pilot training followed at the Militärflieger-Schule in Köslin. After passing all exams, Braun was transferred to Armee-Flug-Park 2. First commanded to the Bavarian Feldflieger-Abteilung 1 on 28 September 1915, he was officially assigned to the Abteilung on 14 November 1915. On 20 November 1915, he was awarded the Bavarian Badge for Military Aircraft Pilots.

His promotion to Leutnant of the Reserve of the Fliegertruppe took place on 23 April 1916. Around six months later, on 4 October 1916, he was transferred to Jagdstaffel 1. He was with this unit until 17 January 1917, when he was transferred to the Bavarian Flieger-Ersatz-Abteilung at Schleißheim. Serving as a flight instructor at Military Flying School 3 in Fürth, Flying School 2

Above: Hanns Braun (right) as silver and bronze medalist at the Olympic Games in Stockholm 1912.

in Lachen-Speyerdorf, and from June 1918 at Flying School 6 in Bamberg, he was able to pass on his experience to the young flight students assigned to him.

He was involved in this activity until shortly before the end of the war. On 15 September 1918, he was transferred to Royal Bavarian Jagdstaffel 34.

On October 9, 1918, at 8:00 a.m., Jagstaffel 34 was engaged in aerial combat with British Sopwith Dolphins, during which Uffz. Franz Ulm was able to shoot down one of these opponents near Maretz. The Staffel returned to the airfield for refueling and ammunition resupply and took off again at 10:00 a.m. in order to resume operations at the front. The Fokker D VIIs of Jagdstaffel 34 were over the front north of St. Quentin at about 11:00 when Uffz. Franz Ulm rammed Lt. Hanns Braun's Fokker D VII from behind at an altitude of about 5,000 meters. Hanns

Braun's plane immediately went into a tight spin, and soon after it hit the ground near the village of Croix Fonsomme. The celebrated sportsman and well-known artist Hanns Braun was dead. Uffz. Franz Ulm was able to land his damaged Fokker D VII inside the German lines and remained unharmed.[8] The reason for Franz Ulm's flight maneuver which caused the accident remains a mystery because there was, according to the flight log of Oblt. Greim, no contact with any hostile aircraft during that mission.

In the flight log of Staffelführer Oblt. Robert Greim is written:

"Fokker D VII 833/18, 10:00 to 11:30, clear, Hight: 5 000 m, Busigny.

No hostile air activity.

Lt. Braun collided with Uffz. Ulm. Lt. Braun crashed beyond the lines from 5,000 meters. Impact observed south of Beantreux (Etaves-et-Bocquiaux, author's note) on the southern corner of the forest."[9]

Hanns Braun's grave is not known. In the Kriegsstammliste (war roster) of Bavarian Jagdstaffel 34 there is a letter dated April 14, 1943, addressed to the sports journalist and author Josef Michler. It states:

"There is no report of his grave here. Presumably his body was buried in the collective grave of the final German military cemetery at St. Quentin.[10]

In 2008, Hanns Braun was inducted into the "Hall of Fame" of German sports.[11]

Albatros D I 435/16, Armee-Flug-Park 1, October 1916, Profile 23

Among the first Albatros D I Jagdstaffel 1 received at the end of August/beginning of September 1916 was the Albatros D I D.435/16. The aircraft is a

Above: Three new Albatros D I, including the Albatros D I D.435/16 are ready for pickup at Armee-Flugpark 1.

Albatros D I 435/16, Armee-Flug-Park 1, October 1916, Profile 23

good example of how the paint scheme of one and the same aircraft could evolve during its frontline service. It also shows that a photo of an aircraft is always only a momentary snapshot and how important supplementing testimonies of former Staffel members are.

Albatros D I D.435/16 was an early example of the Albatros D I production series, which carried the military numbers D.422/16–D.471/16. This particular D I was one of a number of aircraft of this type that were delivered to the German 1st Army during September 1916, and it was assigned to Jagdstaffel 1 and 2.[12]

A photo of Albatros D I 435/16, presumably

taken at the Armee-Flug-Park 1, shows the aircraft as delivered from the factory. The fuselage carried the usual factory honey-colored varnish, the metal parts (struts, engine cowling, and propeller hub) are grayish-green, and the rudder is likely painted in one of the camouflage colors applied to the wing upper surfaces, dark green or rust red.

Albatros D I 435/16, Jagdstaffel 1, September 1916, Profile 24

Likely soon after the aircraft arrived at Jagdstaffel 1, the plywood fuselage was treated to a locally applied camouflage pattern. This pattern may well have been applied in light green, dark green and rust red.

Below: The same Albatros D I D.435/16 at a later date after a camouflage paint scheme was applied in the field. The military number on the fin has carefully been repainted in white. (A. Imrie)

Albatros D I 435/16, Jagdstaffel 1, September 1916, Profile 24

These colors were available in sufficient quantity for purposes of repair and maintenance at the Armee-Flug-Parks, as they were used for the camouflage of the upper wings of many aircraft. The rudder is still in the factory finish, as can be seen by the small Albatros company logo, and brown was chosen for the profile. Since the tail fin was now incorporated into the new three-tone camouflage paint scheme, the factory number has been carefully repainted in white. As can be seen on the front of the fuselage, the camouflage paint was applied in a way that left fairly rough edges along the different colors. The fuselage bottom was very likely painted light blue to match the lower surfaces of the wings.

Oblt. Hans Bethge, Albatros D I D.435/16, October/November 1916, Profile 25

This photo shows the aircraft at a time later, but again in a different livery. According to Franz Ray's statement to Alex Imrie, Oblt. Bethge used a Mercedes star as his personal marking.[13] The previous camouflage paint on the fuselage was now covered by a single solid color. The rudder and tail fin have been painted over in a light color, probably light blue. The Mercedes star was painted in white on the fuselage as a personal insignia.

Hans Bethge retained the Mercedes star as his personal marking on his Albatros D III when he later took command of Jagdstaffel 30. Likewise, two of his aircraft had the vertical stabilizer painted in blue.[14] For this reason, the color of the fuselage is also shown here as blue. As Otto Fuchs, former member of Jagdstaffel 30 rcalled, Hans Bethge was a great fan of Mercedes cars and engines. It seems logical that this enthusiasm originated with this particular Albatros D I flown by him and powered by Mercedes engines.[15]

This aircraft serves as an example that already by the fall of 1916 the fuselage of a fighter aircraft could be painted in different colors over a short period of time.

Hans Bethge was born in Berlin on 6 December 1890 as the son of Kapitänleutnant Wilhelm Bethge. He spent his youth in Friedrichshafen on the beautiful Lake Constance. Since he was barred from joining the navy because of his nearsightedness, he joined the Eisenbahnregiment 1 in Berlin as a Fahnenjunker and received his officer's commission in 1912. At the beginning of the war, he went into the field with this regiment. The tasks of his unit included the expansion and repair of the rail network, as well as tunnel construction and blasting bridges. After recovering from an injury, he enlisted in the Fliegertruppe in 1915. In May of that year, he began his training as a pilot with Flieger-Ersatz-Abteilung 4 in Posen and at the Geschwaderschule Döberitz. In November 1915 he was transferred to the Brieftauben-Abteilung Ostende, later Kampfgeschwader 1, and subsequently flew on the Verdun Front and the Somme. In order to be able to counter enemy fighters, Kagohl I received a pair of single-seat Fokker monoplanes, one of which was piloted by Hans Bethge. On 12 June 1916, he was assigned to Kampfeinsitzer-Kommando-Bertincourt, also named Kampfeinsitzer-Kommando Nord, and attached to Feldflieger-Abteilung 32.

On 22 August 1916, Jagdstaffel 1 was formed from the flying personnel of the Kommando. In the same month he was able to achieve his first two aerial victories, for which he was awarded the E.K. I and the honorary cup "Dem Sieger im Luftkampf" on 12 September. After his third aerial victory, he was appointed Staffelführer of the newly formed Jagdstaffel 30 in the area of operations of the 6th

Above: The Albatros D I 435/16 when flown by Oblt. Hans Bethge. The fuselage has now been repainted in a uniform color, interpreted as blue. The personal marking applied to the aircraft was the Mercedes star.

Right: Magazine advertisement of the Mercedes - Daimler Motoren company in 1918, showing the style of the logo that was then in use. This served as the pattern for the marking on Hans Bethge's aircraft.

Army on 15 January 1917. In September, when the Jagdstaffeln in the army section were combined into one Jagdgruppe, he became its leader in addition to his duties as Staffelführer of Jagdstaffel 30. He was a very popular Staffelführer and called "Papa (Daddy) Bethge" by his pilots. Around the same time he was awarded the Knight's Cross of the Royal House of Hohenzollern. He was submitted for the Pour le Mérite after scoring his 20th aerial victory on 10 March 10, 1918, but fate prevented him from receiving this decoration. On 17 March 1918, he took off on a frontline flight with 4 aircraft of his Jagdstaffel 30. In the area southwest of Roulers, an aerial fight occurred with Airco D.H. 4 two-seaters, which presumably belonged to 57th Squadron RFC. In the course of the dogfight, Hans Bethge crashed, having received fatal hits. His body was transferred to his hometown Berlin and he was buried there.[16]

Offz.Stv. Wilhelm Cymera, Albatros D II D.1723/16, December 1916, Profile 26

The fuselage of his Albatros D II has the honey-yellow factory varnish of Albatros Company. As a personal insignia, the aircraft was marked with a "C", in the "Sütterlinschrift" used mainly in Prussia, but also in other parts of the German

Mercedes
Daimler-
Motoren-
Gesellschaft
Stuttgart-
Untertürkheim

Ausstellungs- und Verkaufsräume in Berlin NW7

Above: Oblt. Hans Bethge, in his quarters as squadron leader of Jagdstaffel 30 in Phalempin. "Papa Bethge," as he was respectfully called by his pilots, was a very popular Staffelführer.

Empire during World War I.[17] The color of the letter was interpreted as red for better visibility. A variation of the color profile also shows a black "C", but this may have been somewhat difficult to see in aerial combat.

The photo of Albatros D II D.1723/17 was taken on 26 December 1916. A group of B.E.2d of 5th Sqn. RFC, intending to attack Vaulx-Vraucourt, was intercepted by aircraft of Jagdstaffeln 1 and 5. In the ensuing dogfight, B.E. 2c 4498, piloted by 2/Lt. H E Arnold, was shot down by Oblt. Hans Bethge. It was his 3rd aerial victory. Vzfw. Paul Bona and Offz.Stv. Wilhelm Cymera had meanwhile attacked B.E.2d 6254 and forced it to land at Sapignies. The British pilot, 2/Lt. F N Insoll, had taken off without an observer in order to be able to carry a heavier bomb load than usual. He was captured by the Germans without injury. Offz.Stv. Cymera landed next to

Oblt. Hans Bethge, Albatros D I D.435/16, October/November 1916, Profile 25

Above: Albatros D II D.1723/16 with Offiz.Stellv. Wilhelm Cymera after a dogfight on 26 December 1916. As a personal marking the aircraft carries a "C" for Cymera, applied in Sütterlin lettering, on the fuselage. (R. Absmeier)

Right: The letter "C" in Sütterlin lettering.

B.E2d 6254, where a photographer had arrived just in time to photograph both the landed British aircraft and Offz.Stv. Cymera's Albatros D II. The aerial victory, however, was eventually awarded to Vzfw. Paul Bona, presumably by draw.[18]

Offz.Stellv. Wilhelm Cymera was born on 5 October 1885, in Elberfeld, now part of the city of Wuppertal. Only fragmentary details concerning his military career are available. In 1916 he flew with a Kampfstaffel in Kagohl 1 on the Somme and achieved his 1st aerial victory. On 22 August, he was shot down in a dogfight near Maurepas, but was

Right: Portrait photo of Offz.Stellv. Wilhelm Cymera.

D.1723/16

**Offz.Stv. Wilhelm Cymera,
Albatros D II D.1723/16,
December 1916, Profile 26**

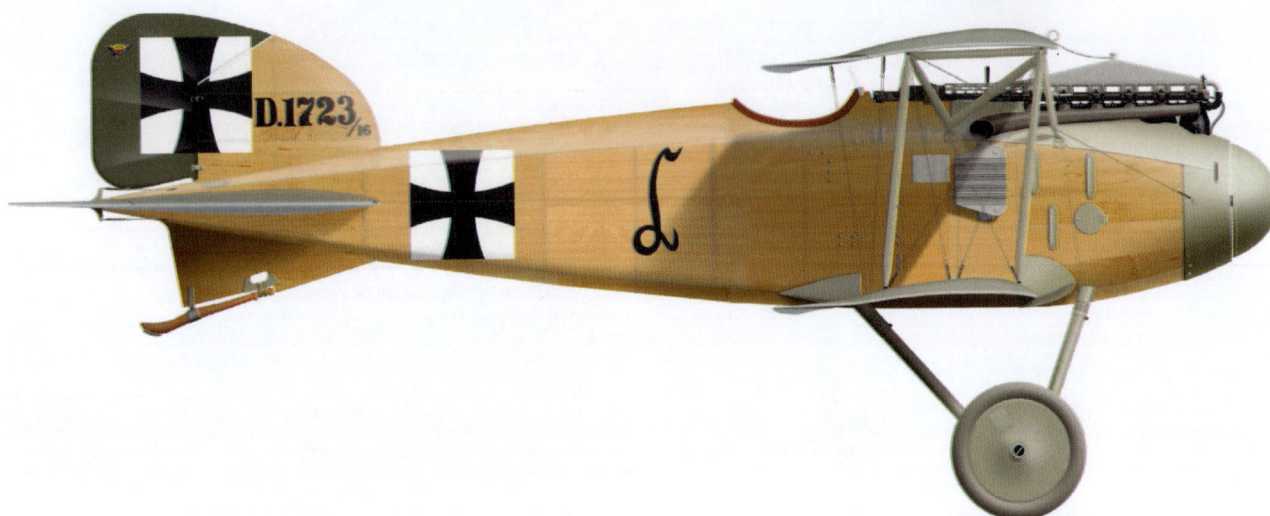

D.1723/16

able to bring down his badly damaged Roland C II in a successful emergency landing, and he remained unhurt. His observer Leutnant Hans Becker had already been mortally wounded in the air. Shortly thereafter, he enlisted as fighter pilot and, after a retraining was transferred to Jagdstaffel 1 around October 1916. Here he scored two aerial victories within a few days in March 1917. After the Staffel was transferred to the Champagne, he was again successful twice in the first days of May. On 9 May 1917, during a frontline flight, he was mistakenly attacked by an Albatros D III of another Staffel and, already fatally hit by the first shots, crashed near Chamouille, his aircraft breaking up in the air.

Lt. Raimund Armbrecht, Albatros D II, February 1917, Profile 27

His Albatros D II had the usual "honey" colored fuselage. His personal insignia was a black "A" on a white rectangle, which appeared on both sides of the fuselage between the cross and the cockpit.

Raimund Armbrecht was born in Leipzig on 11 April 1888. He enlisted as a war volunteer with the Ersatzabteilung Kraftfahr- Bataillon (Motorist-Replacement-Battalion) on 27 August 1914 and was transferred to Kraftwagen-Kolonne (Motor Vehicle Column) 7, Reservekompanie Schöneberg less than four weeks later. On 12 May 1915, he was transferred to Reserve Infantry Regiment 39, 3rd Company, where he was promoted to Unteroffizier on 29 May 1915, Vizefeldwebel on 20 June 1915, and Leutnant on 30 July 1915. He enlisted in the Fliegertruppe and on 14 August 1915 he was assigned to Flieger-Ersatz-Abteilung 2 for pilot training. On 9 November 1915, he joined Flieger-Ersatz-Abteilung 8 and received additional training from 11 November 1915 to 30 March 1916, at Militärfliegerschule Köslin. From there he was transferred to Versuchsflugpark "West" on 13 May 1916 and finally to Kampfstaffel S.3 on 5 July 1916.

On 9 February 1917, he was transferred to Jagdstaffel 1, where he achieved his first aerial victory on 6 September 1917, followed by his second

Right: Lt. Raimund Armbecht sits on the horizontal tail surfaces of his Albatros D II after returning from an aerial engagement on 4 March 1917. Apparently, the left elevator control cable has been severed, and freshly patched bullet holes on the horizontal control surfaces.

Left: Lt. Raimund Armbrecht, seen on the left, in front of his Albatros D II D.1743/16, at Proville airfield in February 1917. At this time a black "A" applied onto a white rectangle served as his personal marking.

Lt. Raimund Armbrecht, Albatros D II, February 1917, Profile 27

While his mechanics pose in front of his Albatros, Lt. Raimund Armbecht is seated in the cockpit of his fighter. The photo was likely taken soon after his arrival at Jagdstaffel 1 in early February 1917, and his plane was probably "pre-owned", since by then the D III was in full production. The "Chapelle Crépin" at Proville can be seen in the background of the photo.

on 16 November 1917. From 12 January 1918 he was entrusted with the deputy command of the Staffel and ten days later was appointed Staffelführer. He led the Staffel until 23 August 1918, at which time, having been ill for some time, he was placed at the disposal of the "Inspektion der Flieger" and transferred to Flieger-Ersatz-Abteilung 8. On 3 October 1918, he joined the Flugzeugmeisterei. On 30 January 1919, he was discharged from military service.[19]

In summary, it can be stated that the Jagdstaffel 1 had already begun to apply individual identification markings to their aircraft soon after its formation. At that time there was at least one and probably several aircraft with fuselages painted in different colors used by the Jagdstaffel.

7.2 Royal Prussian Jagdstaffel 2

Staffelführer:

Hptm. Oswald Boelcke 24 August 1916 – 28 August 1916

Oblt. Stefan Kirmaier   29 October 1916 – 22 November 1916

Hptm. Franz Walz   29 November 1916 – 09 June 1917[1]

In 1975, I visited aviation historian Herbert Schulz in Hamburg. He was a member of the first generation of German aviation enthusiasts who studied the history of German Jagdstaffeln in the 1920s and 1930s. After contacting Cross and Cockade USA member Hans-Eberhardt Krüger via the aforementioned publication's address list, Krüger – a Detective Chief Inspector by profession – brought me in contact with Herrn Herbert Schulz.

His photo collection was small, but his great treasure was his documentation of Jagdstaffeln 2 and 5, which he had compiled in the 1920s and 1930s, based on the testimonies of former members of the two Staffeln. As a young man he had come in contact with Pour le Mérite holder Paul Bäumer, who also invited him to a flight over the Hamburg harbor. He also brought him in contact with the "Kameradschaftlichen Vereinigung der Jagdstaffel Boelcke" (Comradeship Association of Jagdstaffel Boelcke). Paul Bäumer even made sure that Herbert Schulz was invited as a guest to the meetings, which were organized by the last Staffelführer Carl Bolle in the 1920s.[2]

Among his records was also a list of the painting schemes of the aircraft of Jagdstaffel Boelcke. He had compiled this in order to be able to build aircraft models of the planes flown by the various pilots of Jagdstaffel Boelcke. As he told me, he used soft lime wood for the fuselage which was easy to carve into the desired shape, and thin plywood for the wings, which was slightly bent over hot steam in order to obtain the desired wing shape. To my deep regret, his airplane models had not survived the ravages of time – I would loved to have seen them.

Above: The Albatros D II prototype D.386/16 at the Albatros factory in Berlin-Johannisthal, circa June/July 1916. When photographed, the aircraft was still undergoing evaluation and had not yet been accepted by IdFlieg, which explains the lack of the military number on the tail fin. This prototype was eventually delivered to Oswald Boelcke on 16 September 1916, and it was destined to become the last plane to be personally used by him.

His records showed that several aircraft of Jagdstaffel 2 had fuselages painted in different colors as early as the fall of 1916, and in addition to that the use of letters as personal markings had also been implemented around this time:

Albatros D I/D II of Jasta 2:

Fuselage painted in red
Fuselage painted in yellow
Fuselage painted in blue
Fuselage painted in black
Fuselage painted in green.
Different letters "Co"; "Bü", "K", "W"... [3]

During one of my visits to Harpenden I told Alex Imrie about the documentation compiled by Herbert Schulz, and he confirmed that, to his knowledge, at least two aircraft of the Staffel already had painted fuselages at that time. These were the "pea green" Albatros D I of Lt. Diether Collin and the red Albatros D II of Lt. Manfred von Richthofen. In addition, he confirmed the statements of Mr. Herbert Schulz by stating that he had also come to the conclusion that they were not the only Albatros D I and D II aircraft with painted fuselages in this unit.

The written information of Mr. Schulz and the theory of Alex Imrie was confirmed sometime later by two photo albums: The first of these was the photo album of Clemens Vollert, former observer in Bavarian Feldflieger-Abteilung 6 and Offizier z. b. V. of Jagdstaffel 76. To my great surprise, his estate held a very nice photo album which exclusively contained photos of Jagdstaffel 2, taken in the autumn of 1916.

Years later I was able to copy the photo albums of Erwin Böhme, Kampfgeschwader II, Jagdstaffel Boelcke, and Jagdstaffel 29. The photos in these albums proved that the plywood fuselages of other Albatros D I and D II aircraft flown by Jagdstaffel Boelcke were painted. This can be seen, among other things, by the fact that on these aircraft the metal parts of the engine cowling and the cooling louvres were painted over in the same tone as the rest of the fuselage. In addition, the military number on the vertical stabilizer had also been painted over or, in some cases, the paint was applied around the number, leaving it in its original state.

In contrast, photos of the unpainted Albatros D I and D II show the engine cowling painted and the cooling louvres in the uniform gray-green factory color, which differs in shade of gray from the color of the wooden fuselage. Likewise, on the unpainted Albatros D I and D II, the military number on the tail fin is clearly visible. [4]

Additional highly interesting photos were provided to me by Lance Bronnenkant, which confirmed the painted fuselage on some Jasta 2 aircraft, as well as photos made available by aviation historians Rainer Absmeier and Johann Ryheul, and the analysis of these photos by Jörn Leckscheid. [5]

Putting all of these photos in connection with the information recorded by Herbert Schulz results in a

Above: A contemporary photo caption claimed that this picture depicts Hptm. Oswald Boelcke preparing to board his Albatros D II D.386/16 just before what would become his final frontline flight on 28 October 1916. However, this caption cannot be correct because…

Below: Hptm. Oswald Boelcke steps out of his Albatros D II D.386/16 after his 36th aerial victory on 16 October. In this photo, the military number is fully visible, and the mottled camouflage on the left lower wing shows well.

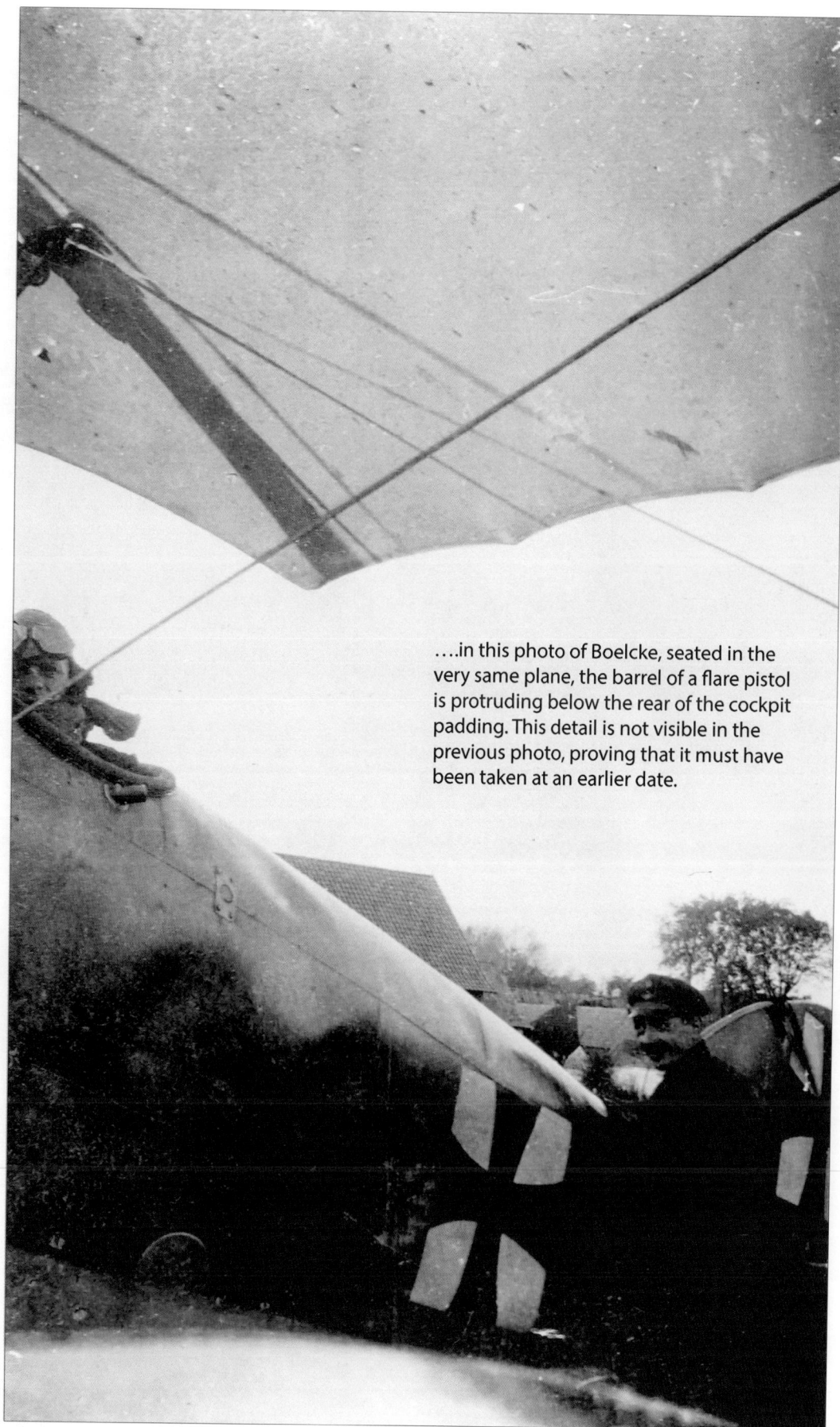

....in this photo of Boelcke, seated in the very same plane, the barrel of a flare pistol is protruding below the rear of the cockpit padding. This detail is not visible in the previous photo, proving that it must have been taken at an earlier date.

Hptm. Oswald Boelcke, Albatros D II D.386/16,
September 1916, Profiles 28 and 28a (top view)

rather more colorful picture of the fighter planes of Jagdstaffel 2 than has been previously thought, even at this relatively early stage of its existence.

Hptm. Oswald Boelcke, Albatros D II D.386/16, September 1916, Profiles 28 and 28a (top view)

The Albatros fighter supplied to Oswald Boelcke for his personal use in Jasta 2 was a very special aircraft indeed, for it actually was the prototype of the Albatros D II series. The other Albatros fighters supplied to his unit at the time were of the D I type, which in itself was a new type at the front. A well-known photograph of this D II taken at Berlin-Johannisthal around late July 1916 depicts the plane still without the military serial applied to the tail fin. This fact indicates that at the time the aircraft had not yet been accepted by the military authorities when the photo at Johannisthal was taken

The fact alone that the prototype of the very latest fighter design produced by the German aviation industry was made available to Oswald Boelcke is ample proof of the unparalleled status he held in the Fliegertruppe in the late summer of 1916. When it was delivered to him, the aircraft did carry the

military number D.386/16 on the tail fin. A study of all the available photos of this plane taken during its frontline service of roughly six weeks reveals that Oswald Boelcke's Albatros D II D.386/16 carried no personal markings. However, the extensive analysis of existing photos highlighted some special features not seen on production examples of the D II. This especially concerns the camouflage pattern applied to the wings, which provide a broader view of this aircraft, also for the modeler.

The fact that Oswald Boelcke was flying the only Albatros D II at the front at the time may have been the reason why he chose not to have a personal marking. Even so, the aircraft was easy recognizable.

Oswald Boelcke was born on 19 May 1891 in Giebichenstein near Halle an der Saale. He was the first "great ace" of the German Fliegertruppe and the preeminent fighter pilot and tactician in the first half of the war. Fixated on a military career at an early age, he joined the Prussian Cadet Corps in 1911 and initially served with Telegraph Battalion No. 3 in Koblenz.

His interest in flying arose during the great

Above: Taken on the same occasion, Boelcke is now surrounded by members of his Staffel who are eager to hear his account of the fight. Meanwhile, Boelcke is still cleaning the powder stains off his face. The mottled application of the wing camouflage is also visible here.

Above: A rare view of the right side of Hptm. Oswald Boelcke and his D.386/16. A flare cartridge holder has been mounted below the cockpit padding. An important identifying detail of the D II prototype shows well in this photo: the rectangular metal joint, seen behind the cockpit cutout, was not a feature seen on production examples of the Albatros D II. This can be noted on both sides of D.386/16.

autumn maneuvers of 1913, and the following year he transferred to the Fliegertruppe as a Leutnant. He received his pilot training at the Halberstadt Flying School beginning in June 1914. Transferred to Flieger-Ersatz-Abteilung 3 at Darmstadt with the outbreak of war, he was awarded his pilot's badge on August 15 and transferred to Feldflieger-Abteilung 13 via Etappen-Flug-Park 4 on 1 September. For the rest of the year, he flew his missions with his older brother Wilhelm as an observer. He was awarded the E.K. II in October 1914 and on 18 January 1915 the E.K. I followed.

On 24 April 1915, he was transferred to Feldflieger-Abteilung 62, where he achieved his 1st aerial victory in July. At the end of the month the Abteilung was assigned one of the first Fokker monoplanes. Being one of the of the most experienced and aggressive pilots flying with the Abteilung, the aircraft was assigned to him. On 19 August, he scored his first aerial victory as a fighter pilot with the Fokker. After his third aerial victory, he was transferred to Brieftauben-Abteilung-Metz

(later Kagohl 2) on 19 September as a single-seater pilot, where he achieved three more kills by the end of October. For this he was awarded the Knight's Cross of the Royal House Order of Hohenzollern on 1 November 1915.

Reassigned to Feldflieger-Abteilung 62 at Douai on 11 December 1915, he received the honorary cup "Dem Sieger im Luftkampf" on Christmas Eve 1915. In January 1916, he achieved three more kills; after his 8th aerial victory on 12 January, he became the first fighter pilot (together with Max Immelmann) to be awarded the Order Pour le Mérite. On 20 January, 1916, he was transferred as a single-seater pilot to Artillerieflieger-Abteilung 203 on the Verdun Front, where he took command of Fokkerstaffel Sivry on 11 March. By May 1916, he had increased his number of aerial victories to 18, placing him at the top of the list of all German fighter pilots; his preferred promotion to Hauptmann followed on 21 May 1916.

After his 19th aerial victory and the death of Max Immelmann in June 1916, he had to relinquish command of his Fokkerstaffel at the beginning

Above: Albatros D I (very likely D.385/16) fitted with a water extension tank in front of the first cylinder of the engine, was later flown by Lt. Collin. He apparently had the fuselage painted yellow, the bottom of the fuselage painted light blue, and a black "Co" outlined in white was applied to the fuselage sides as well. The sheen of the glossy yellow paint is apparent in the photo. (G. VanWyngarden)

of July and was sent on an inspection tour to the Eastern Front and Turkey. During this inspection trip, he wrote a memorandum in which he outlined his ideas regarding the organization and tactical use of fighter aircraft. This memorandum led to a complete rethinking in the Supreme Army Command regarding the use of single-seat fighters and later gave him the unofficial title of

"Lehrmeister der deutschen Jagdflieger" (Master teacher of the German fighter pilots).

In mid-August 1916 – at the height of the summer battle – he was recalled to France from his inspection trip in Turkey and appointed Staffelführer of the newly formed Jagdstaffel 2 on 27 August. Because of his reputation in the Fliegertruppe, he was able – to a certain extent – to choose his own

Below: During the repainting of the aircraft, the fin and rudder cross was reduced in size, and the fuselage cross was re-applied a bit further forward. Compare these details to the photo of Vzfw. Reimann seated in the cockpit of the newly-delivered D.385/16 in chapter 4. (J. Ryheul)

Lt. Diether Collin, Albatros D I (likely D.385/16), September/December 1916, Profiles 29 and 29a.

airmen. Thus, he brought unknown pilots into his Staffel, including Manfred von Richthofen, Erwin Boehme, Max Müller, Otto Hoehne and Stephan Kirmaier, all of whom were later to make a name for themselves. In the next two months he achieved extraordinary successes with his Jagdstaffel and was even able to increase his victory tally to 40 by adding 21 more aerial victories to his previous score, making him by far the most successful fighter pilot ever at that time. But only two days after his 40th aerial victory, on 28 October 1916, his stellar flying career came to an abrupt end: in a dogfight with a "Vikkers" squadron the aircraft of his friend Erwin Boehme collided with his Albatros D II. Following an unsuccessful emergency landing attempt due to a broken wing, he fatally crashed from a low altitude at about 5:00 p.m. at the controls of his Albatros D II D. 386/16 between Bapaume and Grevillers.

In his honor, his Jagdstaffel 2 was renamed "Jagdstaffel Boelcke" by Imperial Decree (AKO) on December 17, 1916. This tradition was continued in later times: During the Third Reich, Kampfgeschwader 27 bore the honorific name "Kampfgeschwader Boelcke," and in the new German Bundesluftwaffe, Jagdbombergeschwader 31 bears his name.[6]

Lt. Diether Collin, Albatros D I (likely D.385/16), January/February 1917, Profiles 29 and 29a

This aircraft serves as a striking example of how the paint scheme of an aircraft could change several times during its frontline service!

An important identifying feature of this particular Albatros D I is that a cooling water expansion tank was mounted in front of the front engine cylinder. This was only the case on less than a handful of prototype and pre-production examples of the Albatros D I. Another identifying feature of these pre-production aircraft was the position of the fuselage cross: it was located further forward than on production aircraft. An analysis of available photos of early Albatros D Is at the front proves that D.385/16, showing the same expansion tank in front of the first cylinder as well as the forward position of the fuselage cross, arrived at Jagdstaffel 1 in August 1916 with Vfw . Reimann in the cockpit. In this context, the war diary of Jagdtaffel 2 states:

"Vfw. Reimann brought an Albatros D I from Jagdstaffel 1 on 1 September 1916."[7]

The Albatros D I flown at a later point by Lt. Diether Collin at Jagdstaffel 2 features the same aforementioned rare details. Available photos prove that this expansion tank is not seen on any other Albatros D I of the Staffel, thus it can be assumed that this is Albatros D I D 385/16 formerly used by Jagdstaffel 1 and brought along by Vfw. Reimann.

Initially, the fuselage of the aircraft appeared in the factory varnished plywood "honey-colored" scheme, and it bore no personal identification. Two photos taken later show that the fuselage of the aircraft was painted with a color that is reproduced very dark on the orthochromatic photos. The fuselage bottom appears in a light gray in these photos. The monogram "Co" is painted on the fuselage in black with a narrow white outline.

Franz Piechulek, a former member of Jagdstaffel 56, told me that when Lt. Diether Collin took over Jagdstaffel 56 in April 1918 as Staffelführer, he changed the Staffel marking. The noses of all aircraft were now painted yellow. His personal aircraft

Lt. Diether Collin, Albatros D I (likely D.385/16), January 1917, Profiles 30 and 30a.

additionally had a yellow band around the fuselage and a yellow tail.[8] The choice of the color yellow may be related to the fact that Diether Collin was a Silesian. The national colors of Silesia are white and yellow. The color yellow is considered the national color of Silesia, just as the color blue is for the Bavarians

Since Diether Collin introduced the color yellow as the Staffel marking at Jagdstaffel 56 and also used it as the personal identification of his aircraft, it seems logical that the fuselage of his aircraft was painted yellow at that time. Herbert Schulz's records also mentioned an aircraft with a yellow fuselage. In this context, it must be pointed out that yellow tones with a higher red/orange content are rendered as very dark, or black, on orthochromatic film.[9] The light underside of the fuselage was probably painted in light blue since this was the factory-applied color of the lower surfaces of the wings. As the photos further show, at this time the wing upper surfaces have the two-tone rust red and dark green camouflage paint. The Iron Crosses are applied on white squares, and the fuselage cross is placed further forward on this aircraft than on all other known Albatros D I fighters operated by Jagdstaffel 2.

In this context, an air combat report of 60 Squadron RFC dated 27 December 1916, referring to an aerial engagement with fighter planes of Jagdstaffel 2 is available, which also confirms the statements made by Herbert Schulz and Franz Piechulek:

> "Combat in the Air
> 2/Lt. Gilles
> No 60th Squadron
> 27.12.16

While on patrol I saw some F.E's and De Havillands attacked by several hostile aircraft. After a little turning I dived on to 1 hostile aircraft and fired. He turned on his right side and disappeared from my view but was to flatten out and fly east. *Remarks on hostile machine: Type, armament, speed etc. 1 large, single seater tractor machine, gun firing through the propeller, and* **yellow in color.**"[10]

It is interesting to note that the reference here is explicitly to a yellow German aircraft, whereas the plywood-covered varnished fuselages of the Albatros D I/D II are regularly referred to as "sand-colored" in British air combat reports.

Lt. Diether Collin, Albatros D I (likely D.385/16), January 1917, Profiles 30 and 30a

Photos taken at a later date show the machine in a very light gray. The formerly yellow Albatros D I was now painted "pea green" at this time, as can be seen from the records by Adolf von Tutschek.[11] The personal "Co" monogram also underwent alteration. The coloring of the "Co" has been reversed, now appearing in white with a black outline.

Photos from the collections of Lance Bronnenkant and Rainer Absmeier show that the wing tops were now also painted a solid light green. The white squares that bordered the crosses on the upper wing deck have been painted over so that only a white border of the Iron Cross remains. This light green paint has already peeled off slightly on the lower left corner of the right cross. Since there are no two different camouflage colors discernible on the upper surfaces of the lower wings, it is clear that these were also painted pea green. The cross on the tail has now been enlarged again and the complete rudder is also painted white. At the top right, a strip of the white paint has peeled off, and below that you can see the dark paint from the earlier photos again. As a further modification, a tachometer has now been added to the front wing strut on the right side.

Since Jagdstaffel 2 was already equipped with Albatros D IIs at this time, it is unlikely that this aircraft was still Lt. Collin's personal aircraft. At this point, the aircraft was probably only retained as a reserve aircraft.

Diether Collin was born on 17 February 1893 in Liegnitz in Lower Silesia. There are only few data about the beginnings of his military career. After graduating from Gymnasium (high school), he took up an officer's career. Initially in the artillery, he transferred to the Fliegertruppe in the summer of 1914. Around October 1916, he joined Jagdstaffel 2 on the Somme, where he achieved his first two aerial victories in the same year. On 21 February 1917, he was transferred to Jagdstaffel 22 in Champagne via Armee-Flug-Park 7. After his 5th kill, he was severely wounded in aerial combat before Verdun on 6 September 1917 and admitted to hospital.

Placed at the disposal of the Inspektion der Fliegertruppe on 30 September, he returned to Jagdstaffel 22 on 4 March 1918, having recovered from his wounds. At the end of the month, he was able to achieve his 6th aerial victory. On 16 April 1918 he was appointed Staffelführer of Jagdstaffel 56. Here he was able to double his number of aerial victories within a quarter of a year.

On the morning of 13 August 1918, he was wounded by a shot in the calf in a dogfight with a Belgian fighter, which he surprised during a balloon attack. Nevertheless, he still managed to bring his aircraft down in a smooth emergency landing. Although his wound was not serious in and of itself, he died late that evening in the hospital at Bailleul

Above: The combat report filed by 2/Lt. G. A. Gilles of No. 60. Sqn. dated 27 December, describes aerial combat against a yellow German aircraft of Jagdstaffel 2. This was very likely the Albatros D I of Lt. Collin.

Above: The flag of Silesia; the yellow of the flag may have inspired the choice of yellow as the identifying color on Diether Collin's aircraft.

from phosphorus poisoning caused by the incendiary ammunition used by his enemy in the balloon attack.[12]

Above: Lt. Diether Collin, pilot of Jagdstaffel 2 and later Staffelführer of Jagdstaffel 56. (A. Imrie)

Prinz Friedrich-Karl von Preußen, Albatros D I (likely D.385/16), Fliegerabteilung (A) 258 (formerly flown by Lt. Diether Collin), March 1917, Profile 31

The Albatros D I was still in service at the front in March 1917, around seven months after its arrival at Jagdstaffel 2. At that time, however, the aircraft was flown by Prince Friedrich-Karl von Preußen, deputy leader of the neighboring Fliegerabteilung (A) 258. It is not known when exactly he took over this aircraft. The "Co" monogram was now painted over, its place being taken by a white skull with crossed bones on a red background. The same marking had also been applied to the propeller spinner.[13]

The skull with the red background was a reference to his parent unit the 1st Leib-Husaren Regiment. This unit had a black fur cap (officers wore a white fur cap) with a red Kolpak, or a black service cap with the skull with crossed bones and a red cap band. In addition, the saber pouch was also red. The 2nd Leib-Husaren Regiment had a white Kolpak, a black cap band, and a black saber pouch to distinguish them.[14]

Above: Yet another paint scheme was applied to Lt. Collin's Albatros D I at a later date. The fuselage is now painted light green, which was reported as "pea green" by contemporary eyewitnesses. Standing at the rear of the aircraft is Prince Friedrich-Karl von Preussen, prepares for the next flight. The photo was perhaps taken at the day when he takes over the aircraft end of January/ February 1917. (A. Imrie)

Left: Another photo of the light green Albatros D I shows that the upper surfaces of both wings have been painted over in green as well, leaving a thick white border around the wing crosses. The tail cross has again been enlarged, and the rudder was now painted white. The colors of the "Co" monogram have been inverted during the repaint, now being white with a narrow black border. (R. Absmeier)

Due to their tradition, the Black Hussars were the most prestigious hussar regiments of the Kingdom of Prussia. On 9 August 1741, King Frederick II founded the 5th Hussar regiment of the Prussian Army, which was initially called the Regiment of Black Hussars. In 1804 the Hussar regiment received the name "von Prittwitz". After the lost battles of Jena and Auerstedt against Napoleon in 1806, it was the only hussar regiment that survived the fall of the Prussian army in full strength and was therefore held in special royal favor. On 20 December 1808, the regiment was divided into the 1st and 2nd Leibhusaren-Regiments. For members of this unit, it was an honor, almost a duty, to wear the emblem of

their traditional regiment on the plane.

On 21 March 1917, a patrol of Jagdstaffel Boelcke spotted the Albatros D I flown by Prince Friedrich-Karl von Preußen during a front line flight, as reported by Oblt. Adolf Ritter von Tutschek, pilot of Jagdstaffel Boelcke:

"It was 21 March 1917, in the afternoon, when a group of Jagdstaffel Boelcke roared off in weather where you could fish in the murk, that is, the clouds hung at 500–600 meters. Sequence (of events): we spotted a hole in the clouds above Cambrai and spiraled up between high cloud mountains. Soon we were through; below us the sea of clouds, and above us the magnificent blue sky. Lieutenant Voss

Right: The Albatros D I of Prince Friedrich-Karl von Preußen was eventually personalized by the application of the emblem of the 1st Leibhusaren-Regiment, a white "skull & crossbones" emblem applied onto a red background. This covered the previous "Co" on the fuselage sides, and it was applied to the spinner, which was now painted white. (L. Bronnekant)

Above: The Albatros D I of Prince Friedrich-Karl von Preußen after being recovered following his emergency landing on British territory. The dark area at the lower wing root is apparently a remainder of the previous yellow fuselage. (A. Imrie)

Prinz Friedrich-Karl von Preußen, Albatros D I (likely D.385/16) , Fliegerabteilung (A) 258 (formerly flown by Lt. Diether Collin), March 1917, Profile 31.

flew in front with the leader plane, and we followed close behind. As I looked around, I suddenly saw a strange airplane approaching us from the left. I soon recognized it as the green Albatros D I of Prince Friedrich Karl.

The prince joined us and, waving his hand merrily, flew up to everyone. He was in charge of a Feldflieger-Abteilung but was to join our Jagdstaffel in the very near future, and for the time being he flew himself in on an Albatros single-seater of our Staffel. On both sides of the airplane and on the propeller canopy he had a large skull painted as his marking.

The Prince often visited us. We were all very fond of him. He was a sportsman in the best sense of the word, and this also showed in the way he flew... Over Arras we spiraled down steeply to possibly intercept an artillery plane. There was none, however, and so we flew south along the Siegfried position we had just taken, close to the lower edge of the clouds. It was clear to all of us that we could not engage in a real cornering fight in the prevailing easterly wind, which would immediately drive us off to the west. Nevertheless, our leader plane, Lt. Voss, suddenly turned over Lagnicourt toward some lattice tails flying around at the same altitude, also close to the lower edge of the clouds. We immediately pushed after them. The dance began. I got behind an Englishman, but he immediately moved into the clouds. Voss had a similar experience. As I looked around, I noticed the Prince's green Albatros, already very close to the ground, descending in the steepest spirals, closely followed by a Vickers lattice tail. While Friedrich-Karl was about to finish off an Englishman, a second one had gotten on his neck and shot up his engine and fuel tank, so that the only thing left to do was

to land immediately. Unable to help, and trembling with excitement, we had to watch as he landed smoothly barely two kilometers from our lines, climbed out, and, running away toward our lines, suddenly collapsed. An Englishman's bullet, as it later turned out, had fatally struck him from behind through the abdomen while running toward our own lines."[15]

Friedrich-Karl von Preußen was born on 6 April 1893 in Klein-Glienicke near Berlin. His father Prince Friedrich-Leopold von Preußen was considered the "enfant terrible" of the House of Hohenzollern and had been largely excluded from the social life of the court by Emperor Wilhelm because of his escapades. This led to the fact that both his sons, Friedrich-Siegesmund and Friedrich-Karl, had to enter the cadet school in Naumburg by imperial order in 1903. Friedrich-Karl joined the Garde Regiment zu Fuß No. 1 as a Leutnant after the Cadet Corps. In 1912 he joined the Leib-Hussar Regiment 1, while his brother joined the Leib-Hussar Regiment 2. Friedrich-Karl was a great sportsman and won a bronze medal as a horseman at the 1912 Olympics. At the outbreak of war, he went into the field with his Hussar Regiment. In October 1916 he transferred to the Fliegertruppe and was deputy Staffelführer of Fliegerabteilung (A) 258. While trying to reach the German lines after his emergency landing, he was seriously wounded. Although it looked as if he would recover, he died due to the wounding between 6 and 9 April 1917, in the military hospital at Saint-Étienne-du-Rouvray.[16] He was **not** the Crown Prince of Prussia as stated previously in some publications. The Crown Prince and heir to the throne was Friedrich-Wilhelm von Preußen, son of Wilhelm II.

Right: The Albatros D I of Prince Friedrich-Karl von Preußen was eventually marked with of the emblem of the 1st Leibhusaren-Regiment, a white "skull & crossbones" emblem applied onto a red background. (B. Gray)

Below: The Albatros D I of Prince Friedrich-Karl von Preußen after his being recovered following his emergency landing on British territory. The header tank in front of the first cylinder shows well, and the four rectangular metal joints mounted along the plywood side panels were only mounted in this way on prototypes of the Albatros D I. These details strongly indicate that this aircraft was indeed 385/16. (B. Gray)

Above: Very nice artwork of the Albatros D I flown by Friedrich-Karl von Preußen was presented by Roden Models on their 1/72 plastic kit of the Albatros D I. The skull marking was actually painted on a red square, rather than the black one shown here.

Lt. Erwin Böhme, Albatros D I D.437/16, Jagdstaffel 2, Fall 1916, Profile 32

The aircraft had a wooden fuselage with a "honey yellow" protective varnish. By the time the photo was taken, the aircraft was already several months old, and the plywood fuselage had darkened somewhat. The military number cannot be made out with certainty in the photo, but the most likely number was D.437/16. As a personal insignia, the aircraft wore a "B" painted on both sides of the fuselage. According to the documents and photos in Erwin Böhme's estate, he used a green "B" or a green fuselage band with a white "B" as his personal marking on the aircraft he flew as fighter pilot.

The use of the color green was a reference to his parent unit, the Garde-Jäger Battalion, which had green uniforms.[17] The "Jäger" had a high reputation in the Prussian Army, therefore it was not uncommon that "Jäger" used the green color as personal markings on their aircrafts.[18] As he used the color green as a personal marking on a number of his fighter planes, it is very likely that the "B" on his Albatros D I would also have been green.

Erwin Böhme was born on 29 July 1879 in Holzminden on the river Weser. He was one of the oldest fighter pilots of the First World War. After graduating from the engineering school in Dortmund, he worked as an engineer in Germany, as well as abroad in Switzerland, and in East Africa. Returning to Germany shortly before the beginning of the war, he volunteered for military service in August 1914 despite being 35 years old. Initially with the Garde-Jäger-Battalion, he transferred to the Fliegertruppe the following month at his personal request and was trained as an pilot at the Flieger-Ersatz-Abteilung 1 in Döberitz. Due to his advanced age, he found use as a flight instructor at the Military Flying School in Leipzig-Lindenthal for a year after completing his training before he managed to get to the front. From December 1915 he flew with Kampfstaffel 10 in Kampfgeschwader 2 on the Russian front and was awarded the E.K. II. He was able to achieve his 1st aerial victory in August 1916. At Kagohl 2 he got to know and appreciate Oswald Boelcke during his inspection tour in the summer of 1916. A close friendship grew out of this, which ultimately led to Boelcke arranging for

Prinz Friedrich Karl von Preußen.

5159

Above: Photo of Prince Friedrich-Karl von Preußen as a member of the 1st Leibhusaren-Regiment. (L. Bronnenkant)

Right: The peacetime uniform of the 1st Leibhusaren-Regiment.

Above: The cap of the 1st Leibhusaren-Regiment.

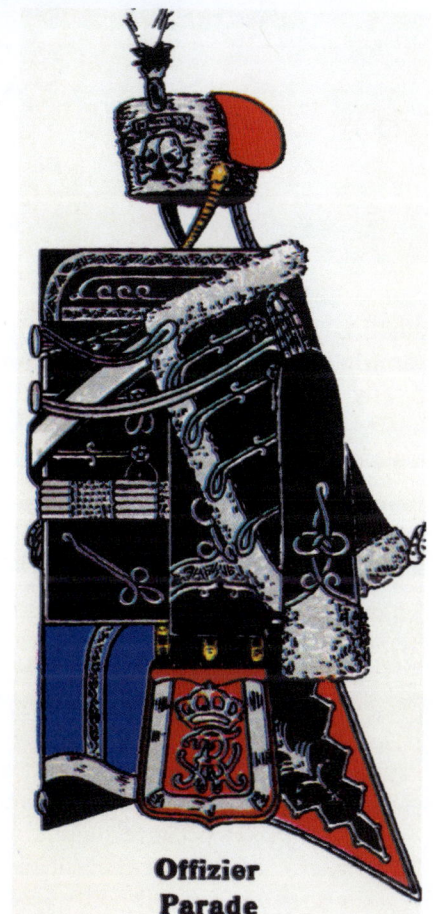

Offizier Parade

Böhme´s transfer to his newly formed Jagdstaffel 2 in September.

He was able to achieve his first aerial victory as a fighter pilot on 17 September during the first frontline flight and was awarded the E.K. I on the same day. On 28 October 1916, fate would have it that he, of all people, collided with the aircraft of his friend and instructor during a dogfight, which caused Boelcke's fatal crash.

Despite this incident, he continued to fly undiminished and increased the number of his aerial victories to 12 by February 1917. On 11 February, he was wounded in aerial combat with Sopwith single-seaters by a shot in the upper arm and had be hospitalized for several weeks. On his sickbed, he received news of the award of the Knight's Cross of the Royal House Order of Hohenzollern on 12 March. After his recovery, he worked as an instructor at the Jagdstaffelschule (fighter pilot school) I in Valenciennes from April 1917, before he was put in charge of Jagdstaffel 29 on 2 July 1917.

With this Staffel he achieved his 13th aerial victory but was soon appointed Staffelführer of Jagdstaffel "Boelcke" on 18 August 1917 as successor to Otto Bernert. Within a very short time he proved to be a worthy "heir" to Oswald Boelcke and led the Staffel, which had been relatively unsuccessful for a prolonged period, to great successes again; he himself achieved 11 aerial victories within three

Above: Erwin Böhme had a green "B" marked on the plywood fuselage of his Albatros D I as his personal marking. This was actually not well visible on the varnished plywood, which darkened over time. On his later fighters he used colored bands as background for his "B" monogram as a consequence. The military number of this plane may be 427/16 or 437/16, although it is not clearly legible in the photo. (J. Ryheul)

Right: The green peacetime uniform with yellow "Gard-braids" of the parent regiment of Erwin Böhme, the Garde Jäger Battalion.

Hauptmann Dienstanzug

months. In recognition of these successes, he was awarded the Order Pour le Mérite on 24 November 1917. However, due to a delay in the transmission of news, he is said to have been unaware of this; it is said that the dispatch with the award message arrived at Jagdstaffel 2 on the afternoon of 29 November, a mere 10 minutes after he had taken off on his last mission. During this flight, he attacked an Armstrong-Whitworth F.K.8 artillery plane over British territory and, hit by its defensive fire, crashed fatally in the area northeast of Zonnebeke.

His remains were buried by the British with military honors at Keerselaerhoek; after the end of the war, he was transferred to Berlin, where he found his final resting place. His correspondence with the nurse Annamaria Brüning and other documents were published in 1930 in the book *"Briefe eines deutschen Jagdfliegers an ein junges Mädchen"* (Letters of a German Fighter Pilot to a Young Girl).[19]

Of all the contacts I have had with relatives of former fighter pilots, the one with the Böhme family

Lt. Erwin Böhme, Albatros D I D.437/16, Jagdstaffel 2, Fall 1916, Profile 32.

Above: Well-known studio portrait of Lt. Erwin Böhme.

has retained a very special place in my memory. In 1982 I inquired at the town hall in Holzminden on the Weser, the birthplace of Erwin Böhme, whether there were still members of his family. Erwin Böhme left no direct descendants and so my hope to find relatives was rather low. To my great surprise I received the address of the two daughters of his brothers, Frau Feldmann and Frau Wölke. I got in touch with the two ladies. It did not take long before I received a very friendly reply to my letter from Mrs. Feldmann with an invitation to visit her. The very next weekend I was on my way to Holzminden. As often, my girlfriend at that time and now wife

Kerstin accompanied me.

Frau Feldmann and her husband still lived in the same house where Erwin Böhme and his brothers had grown up. When I entered the house, I had the feeling that time had stood still since the 1st World War. On the walls in the hall hung trophies of animals Erwin Böhme had shot in Africa and photos from that time. In the living room there were photos of Erwin Böhme and his brother and in a glass case the medals and the cup "Dem Sieger im Luftkampf" were lovingly kept. All this created an atmosphere which I had rarely experienced when visiting relatives of former fighter pilots. When we sat together enjoying coffee and cake, I had the feeling as if Erwin Böhme would come through the door any moment, in full uniform, of course. It was touching and beautiful to see how the family had kept the memory of Erwin Böhme alive. We spent an unforgettable day togehter. In conversation, I also learned that I was the first aviation historian to visit them since the 1930s. The family provided me with the photo albums and all of Erwin Böhme's remaining letters so that I could copy the material. In retrospect, I very much regret that I had not taken photos of the house and the living room.

I stayed in contact with Frau Feldmann for a long time, who complained bitterly in a telephone conversation years later that she was harassed by a German "collector" in the most violent way. After he was **allowed to copy the** photos and letters, **the collector** harassed **the family again and again with the aim to get the original letters, the cup "Der Siege rim Luftkampf" and the medals as a present! I** then contacted the "gentleman" to make it clear to him that he had to stop to harassing the family, which led to a heated argument with this "collector". To my great astonishment, years later an article of this

Above: Albatros D I of Jagdstaffel 2 at Lagnicourt airfield in September or October 1916. The second machine from the right is the red Albatros D I (D.381/16 or D.391/16) of Manfred von Richthofen. The fuselage of the plane on the far right (possibly 427/16) is also painted, on the far left is Lt. Collin's yellow Albatros D I. Positioned just in front of it, with the light-colored fuselage, is Lt. Jürgen Sandel´s D I with the military number D.431/16. The letter "S" has been applied as the personal marking on this machine. (L. Bronnenkant)

"so-called historian" appeared in "Over the Front", about Erwin Böhme, where he presented himself as a "great friend of the Böhme family"...

Lt. Manfred von Richthofen, Albatros D II D.481/16, autumn 1916, Profiles 33, 33a and 33b

According to Herbert Schulz's records, Manfred von Richthofen had painted the fuselage of his Albatros D II / D II red while flying with Jagdstaffel 2. For this reason, we can be quite sure that the aircraft had a fuselage painted red, as it was written in the book "Der rote Kampffliger":

"For some reason, one fine day I got the idea to paint my crate bright red."[20]

The red color was a reverence to his parent unit the Uhlan Regiment Emperor Alexander III of Russia (West Prussian No. 1), which had red cap bands, red collars, and red cuffs.[21] On the nose, a black and white ring was painted, representing the Prussian colors, and together with the red fuselage the combination colours also presented the German flag, black-white-red.

The available photos do not show very clearly if the Iron Cross was overpainted by the red paint or not. For this reason, we show two versions: one with the black Iron Crosses spared by the red paint and one with the Iron Crosses covered by a thin layer of red paint.

In reference to this aircraft, a combat report filed by 60th Squadron RFC dated 27 December 1916 is of also interest:

"Combat in the Air
2/Lt. G.A.H. Pidcocl
Nieuport Dct. No. A208
27.12.16
3.25 p.m.
North of Ficheux
I was on offensive patrol between Adinfer Wood and Neuville and saw 6 hostile aircraft who kept away till we turned for home. I dived and fired 30 rounds at an **all-red hostile scout**, *and soon after a hostile aircraft dived on to three F.E's so I dived and gave the former about a third of a drum, after this he nose-dived and flattened out near the ground.*[22]

Thanks to the support of the former Jagdstaffel 11 pilot Carl-August von Schoenebeck, I established contact with several former fighter pilots of Jagdgeschwader I who, like him, still knew Manfred von Richthofen personally. In their description of his personality, I never gained the impression that Manfred von Richthofen acted on a whim. Whatever he did as a fighter pilot always had a valid reason. Therefore, the question arises whether it was really just a whim to paint his plane so conspicuously or was there another reason behind it? First of all, the conspicuously red-painted aircraft was very easy to spot, both by his Staffel mates in the air and by ground observers alike.

But there may have been another reason: Lt. Manfred von Richthofen was the Staffell's most

Above: Manfred von Richthofen standing next to his red Albatros D II. He military number of this plane most likely was 481/16. (L. Bronnenkant)

Right: Enlargement of another view of Manfred von Richthofen´s red Albatros D II set up for engine maintenance. (L. Bronnenkant)

successful fighter pilot after the death of Hptm. Oswald Boelcke, along with Oblt. Stefan Kirmaier. Jagdstaffel 2 was divided into two groups in the fall of 1916. One group was of course led by the Staffelführer Oblt. Kirmaier. But we have no information on who led the second group.

One reason to paint the entire fuselage red may have been related to the fact that Manfred von Richthofen led a group or "Kette" of fighters with this aircraft. This is indicated by a record of Offz. Stv. Max Müller, pilot in Jagdstaffel 2, in which he writes:

"10. October 1916:

We flew under the command of Oberleutnant Kirmaier on a pursuit flight of six Englishmen in the direction of Cambrai. There we encountered four enemy fighters with which, as we heard later, a **group under Richthofen** *had already fought."*[23]

If you look at the photos in connection with the documents of Herbert Schulz, you can see that the identification of the aircraft at the Jagdstaffel 2, at least with some pilots, was in connection with the color of the cap bands or the color of their peacetime uniforms. This was a "tradition" that was later taken up again and again by fighter pilots for the painting of their aircraft.

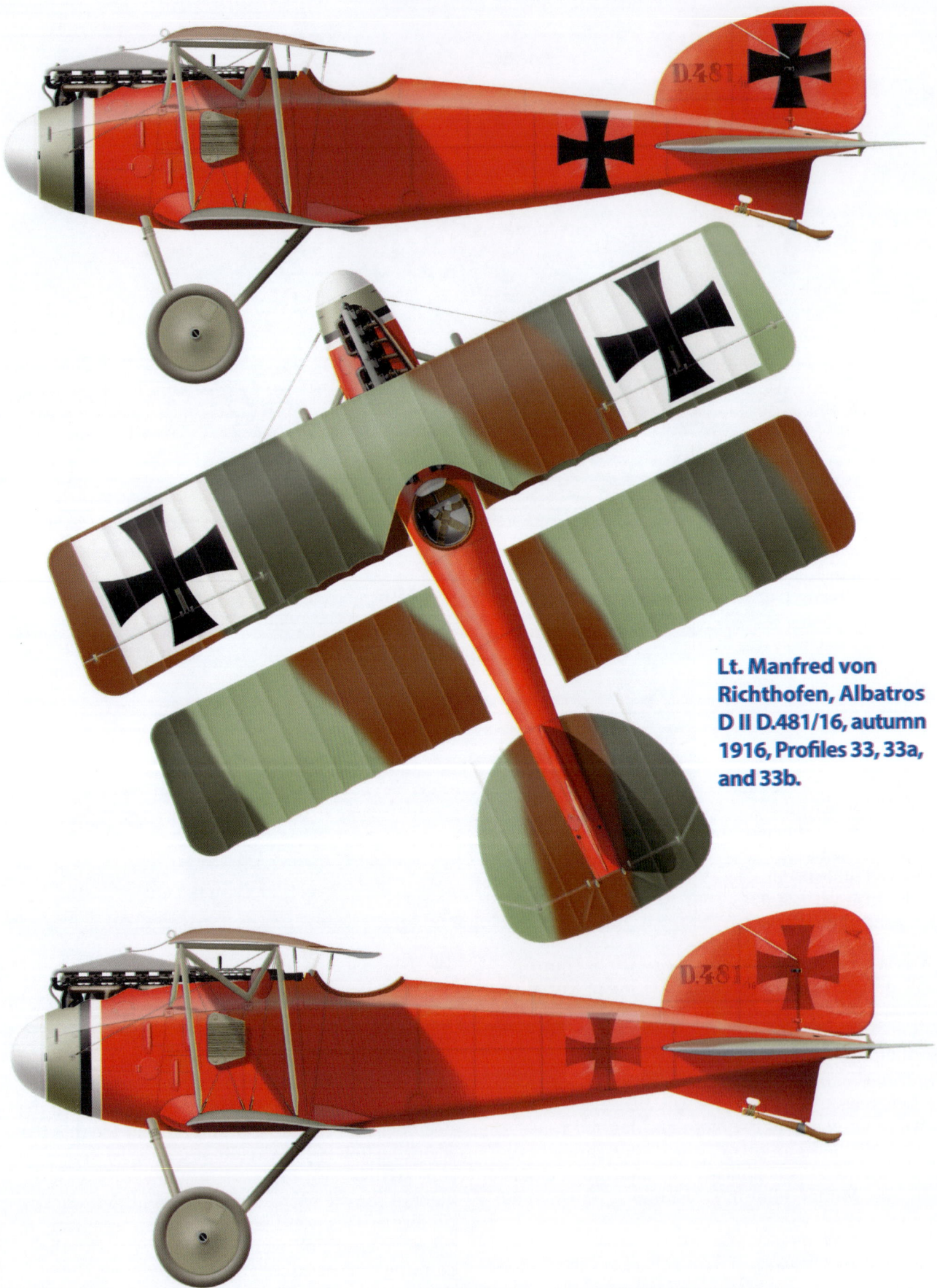

Lt. Manfred von Richthofen, Albatros D II D.481/16, autumn 1916, Profiles 33, 33a, and 33b.

Above: The combat report of 2/Lt. G.A.H. Pidcock of No. 60. Sqn. Also dated 27 December 1916, it describes engaging an all-red German aircraft at around 11.000 feet. This was the Albatros D II flown by Lt. Manfred von Richthofen.

Right: The peacetime uniform of the Uhlan Regiment Emperor Alexander III of Russia (West Prussian) No. 1. The regiment had red collars, red cuffs and a red cap band as identification.

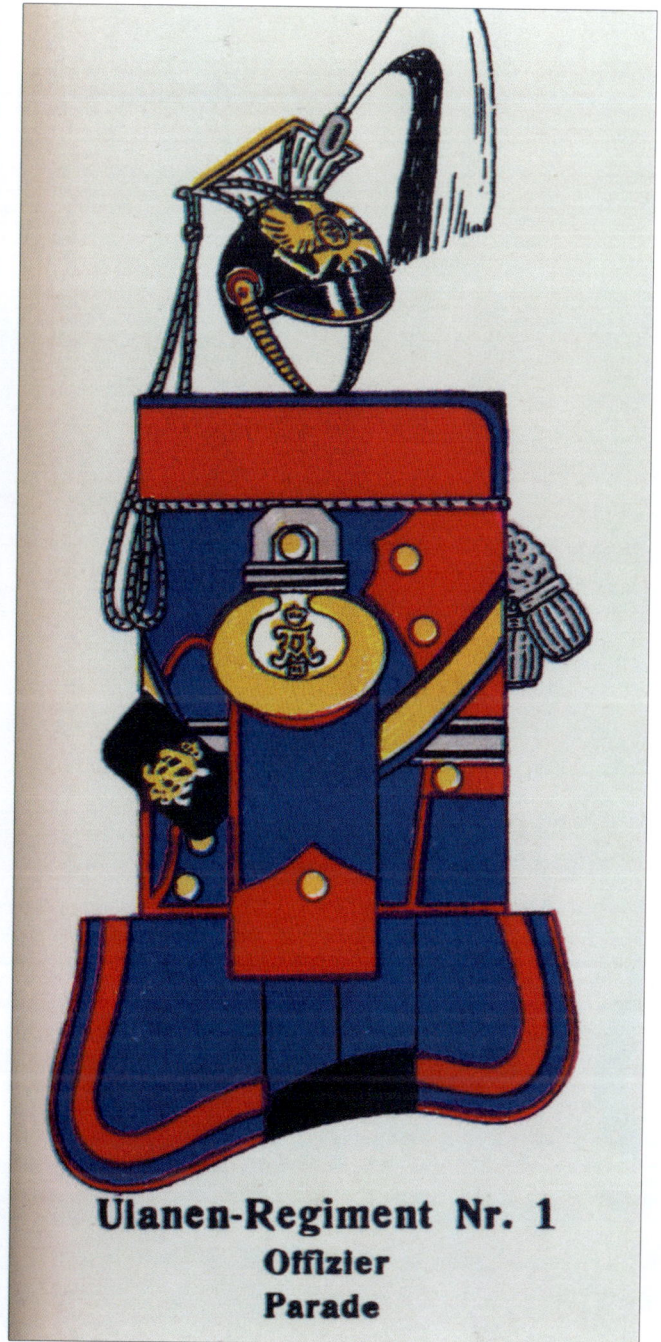

Ulanen-Regiment Nr. 1
Offizier
Parade

7.3 Royal Prussian Jagdstaffel 3

Staffelführer:

Oblt. Hermann Kohze 31 August 1916 – ? September 1918[1]

In terms of available photographic records, Jagdstaffel 3 is one of the most poorly documented units. For the period August 1916 – winter 1916/17 there are unfortunately only 3 photos from the collection of Reinhard Zankl that can be attributed to this period.

These photos were probably taken at Fontaine-Uterte airfield in late 1916. The Staffel was stationed at this airfield from 5 November 1916 to 20 March 1917.[2] The reason for this conclusion is that the Albatros D IIs were aircraft of the first Albatros production batch (D. 472/16–D. 521/16) as well as one aircraft of the second production batch (D.1701/16). These aircraft are supposed to have been delivered to the front in October and November 1916. Also, there is no Albatros D III to be seen,

Above: Albatros D II of Jagdstaffel 3 at Fontaine-Uterte airfield late 1916. Two of the airplanes feature personal markings. It is possible that these were the personal planes of the Staffelführer and the leader of a "Halbstaffel" (half-Staffel). (R. Zankl)

Left: The Albatros D II of Jagdstaffel 3 at Fontaine-Uterte airfield in late 1916 seen from a slightly different perspective. (R. Zankl)

Above: The Albatros D II, likely D.503/16, marked with the light-colored fuselage band, interpreted as white. (R. Zankl)

Left: Enlargement of Albatros D II D.503/16 seen in the first lineup photo of Jagdstaffel 3. (R. Zankl)

Albatros D II D.503/16, March/ April 1917, Profile 34

which would be expected at Jasta 3 in late March 1917 when the unit was stationed at the airfield of Guesnain.[3]

Most of the aircraft are without personal identification or painting. Only two aircraft bear individual identification. One Albatros D II has a bright band around the fuselage, and another Albatros D II has a multicolored band as an insignia. These may be the markings of the leaders of the "Halbstaffeln" (half-units).

Any additional photos of Jasta 3 covering the timespan from the formation of the unit to early 1917 would be very much appreciated indeed! Should these be made available to us, they will be included in the chapter "New results of research" in future volumes in this series of books.

Albatros D II D.503/16, March/ April 1917, Profile 34 (Interpretation)

The aircraft had a "honey" colored fuselage. For the purpose of identification, a fairly wide fuselage band had been painted around the fuselage. This appears white in the photo. Actually, the fuselage band could also be light blue and may have been a reference to the cap band of the pilot's parent unit. We have decided to use a white band for the display. But in any case, this is only speculation. The exact military number of this aircraft is difficult to discern in the photo, and D.503/16 has been chosen as the most likely number.

7.4 Royal Prussian Jagdstaffel 4

Staffelführer:
Oblt. Hans-Joachim Buddecke   25 August 1916 – 15 December 1916
Oblt. Kurt-Bertram von Döring 16 December 1916 – 22 February 1918[1]

The color profiles of the airplanes operated by the Staffel in 1916 are based on the information Alex Imrie obtained from former Staffel-member Alfred Lenz, as well photos in the photo album of the former Staffel member Ernst von Althaus which I was able to photograph in the 1970s, as well as the photos of Hans-Joachim Buddecke, from the collection of Tobias Weber, who kindly made them available to me.

Since its formation on 25 August 1916, the Staffel was located at Roupy airfield in the area of the

German 2nd Army. The available photos show that the Staffel was initially equipped with Halberstadt D III, Fokker D I and at least one Fokker E-Type, likely an E III. Soon, Halberstadt D Vs were delivered to the unit, and for a time the unit seemed to operate a mix of Halberstadt D-types as its main equipment. On 21 December 1916, the Staffel began its transfer to Avillers Airfield in the German 5th Army area, which was completed on the 27th of that month. According to information from the transcript of the war diary, after arriving at the new airfield, the Halberstadt aircraft were given to the Armee-Flug-Park and the Staffel received new Albatros D IIs. This was also confirmed by Alfred Lenz. The photos in Ernst von Althaus' photo album, taken at Avillers Airfield, prove that the unit was fully equipped with the Albatros D II by then.[2]

Left: Oblt. Hans-Joachim von Buddecke in front of his Halberstadt D V, which he himself described as "brown rat".

Above: Jagdstaffel 4 in September 1916, from left: 1. Vzfw. Patermann, 2. Lt. Fust, 3. the Paymaster, 4. Lt. Füger, 5. Lt. Schütz (guest), 6. Oblt. Buddecke, 7. Lt. Bernert, 8. Lt. Stehle, 9. Lt. Malchow, 10. Lt. Kralewsky. The inscription in the lower left corner of this photo reads: "In grateful admiration to our trusted leader, the non-commissioned officers and crew of Jagdstaffel 4".

**Lt. Hans-Joachim von Buddecke,
Halberstadt D V, autumn 1916, Profile 35**

Lt. Hans-Joachim von Buddecke, Halberstadt D V, Autumn 1916, Profile 35

His Halberstadt D V had a blue stripe before and aft of the Iron Cross as personal marking.

The Halberstadt D V of the Staffel had a fabric covering which was described by Oblt. Hans-Joachim Buddecke in his book *El Schahin – Der Jagdfalke* as "**brown**" or "**brown rats**".[3] The color of the bands is not known and was interpreted as blue. See also 8.1 – The colors of the factory covering of the Halberstadt fighter planes.

Hans-Joachim Buddecke was born in Berlin on 22 August 1890. He decided early to become an officer and entered a cadet school in 1904. Promoted to Leutnant in 1910, he left the military in early 1913 and emigrated to the U.S.A., where he first worked in his uncle's automobile company. That same year, he acquired a Nieuport biplane and learned to fly in his spare time. His plan to open his own aircraft factory, however, was thwarted by the outbreak of war in Europe. Aboard a Greek freighter, he returned to Germany and volunteered for the Fliegertruppe. Initially with Flieger-Ersatz-Abteilung 3 in Darmstadt for a short time, he flew with Feldflieger-Abteilung 23 from the beginning of 1915 and was one of the first pilots to fly the new Fokker monoplanes. He achieved his first aerial victory as early as September 1915. By the end of the year, he had added two more, making him one of the most successful fighter pilots in this early phase of the war and earning the E.K. I, the E.K. II and, on 16 October 1915, the Knight's Cross of the Saxon "Militär St. Heinrichs Orden IV Klasse".

In early December 1915, he and a handful of other airmen were sent to the Dardanelles Front to support the Ottman ally, who was hard pressed by the British and French, in response to an official request for help. Officially members of the so-called German Military Mission in Turkey, they were assigned to Ottoman aviation detachments, most of them to Ottoman Fliegerabteilung 6 at Smyrna. After the first Fokkers arrived in Smyrna shortly after Christmas, Buddecke was able to bring down three British planes within a week, whereupon the activities of the British aviators on this front abruptly declined. After a fourth aerial victory that same month, he was awarded the Ottoman Golden Liakat Medal and the Silver and Golden lmtiaz Medals in just a few days. He was also awarded the Knight's Cross of the Royal House Order of Hohenzollern. On 4 April 1916, he was able to achieve his 8th aerial victory and was awarded the Order Pour le Mèrite for this on 14 April, the third fighter pilot after Oswald Boelcke and Max lmmelmann to receive this highly prestigious award. Until August 1916, "EI Shahin" (the hunting falcon) – as he was respectfully called by the Turks – almost single-handedly held the British pilots at bay on the Dardanelles Front, without, however, being able to achieve any further successes, after which he was recalled to Germany.

On 25 August, he took over as Staffelführer of Jagdstaffel 4, newly raised largely from personnel of his old Feldflieger-Abteilung 23 and scored his 9th and 10th aerial victories on the Somme in September 1916. On 15 December 1916, he relinquished the command of the Staffel; as a result of a new request from the Ottoman Emire, he returned to the Bosporus and was assigned to Ottoman Jagdstaffel 5 at his old post in Smyrna. Once again, it was almost his presence alone that forced the British airmen to exercise greater restraint; but only on 30 March 1917, did he achieve two aerial victories.

After almost 13 months in the Ottoman Empire, he returned home at the end of January 1918. On 15

February, he arrived at Jagdstaffel 30, in order to first gain frontline experience under the – compared to the Dardanelles – completely different conditions of the Western Front. Just four days later he was able to achieve his 13th aerial victory.[4]

The former fighter pilot of Jagdstaffel 30, Hans Holthusen remembered the fighting style of Hans-Joachim Buddecke during the joint frontline flights:

"Buddecke went about it with a bravado that made our blood run cold".

After three weeks with Jagdstaffel 30, he joined his old friend Rudolf Berthold on 8 March as deputy Staffelführer of the latter's Jagdstaffel 18. Since Berthold was still recovering from his most recent injury and was not yet fit to fly again, he was to lead the Staffel in the air. But on 10 March 1918, during his first flight with the new Staffel, he met his fate: at about 1:10 p.m. he crashed near Harnes after aerial combat with Sopwith Camels, mortally wounded by a heart shot. His mortal remains were transferred to his hometown Berlin and buried there. His war experiences in France and the Ottoman Empire were published in 1918 in the booklet *El Shahin – der Flug des Jagdfalken* (the Flight of the Hunting Falcon)".[5]

Lt. Ernst von Althaus, Halberstadt D III, autumn 1916, Profile 36

The aircraft had a yellow "A" as a personal insignia on the aircraft, which was clearly visible on the "rat-brown fuselage" of the Halberstadt D III. A black "A", as depicted in some publications, would have been difficult to make out on a background of this color in flight. He must also have had a special affinity for the color "yellow" because, as Staffelführer of Jagdstaffel 10, he introduced this color as Staffel making when he took over this Staffel on 6 July 1917.[6]

Ernst Freiherr von Althaus. His father, Prince zu Bentheim and Steinfurt, belonged to an imperial noble family of Westphalian origin and was adjutant to the Duke of Saxe-Coburg-Gotha. His mother, a well-known stage actress, was of middle-class origin. According to the then prevailing Wilhelmine legislation, the children of such a marriage lost the title of nobility. Due to the high position of the Prince of Bentheim and Steinfurt, the wife received the noble title of a Freifrau von Althaus after the marriage, albeit a much lower one. Thus, he was born as Ernst Freiherr von Althaus.

Ernst von Althaus underwent the usual training in a cadet school at that time and was promoted to Leutnant in 1911 with the Saxon Husaren Regiment

Above: Snapshot of Oblt. Hans-Joachim Buddecke, taken by Ernst von Althaus. According to Alfred Lenz, Buddecke was very popular with officers and crew because of his friendly leadership style

"König Albert" No. 18. However, on 20 February 1914, he suddenly retired from active duty and was transferred to the reserves just as he was scheduled for training as a military pilot with A.E.G. The strict military discipline did not exactly suit his nature.

At the outbreak of the war, he went into the field as a reserve officer with his old regiment in August 1914. As a Hussar he was in his element in the war of movement during the first weeks of the war and quickly drew attention to himself with daring missions for which he was awarded the E.K. II and E.K. I, as well as the Knight's Cross of the Saxon Military Order of St. Heinrich on 27 January 1915.

In May 1915 he switched to flying and was trained as a pilot at Flieger-Ersatz-Abteilung 6 in Großenhain. On 25 July 1915, he was transferred to Feldflieger-Abteilung 23, where he initially flew with various observers in two-seaters. On August 6, he was promoted to Oberleutnant. In October 1915 he received his training as a Fokker pilot at Sandhofen near Mannheim at the school of the Kampfeinsitzer-Abteilung 1.

As a pilot of the Kampfeinsitzer-Kommando Vaux,

Facing Page, Above: A yellow "A" served as the personal identifier for the Halberstadt D III flown by Oblt. Ernst von Althaus. Note the black and white leader's streamer attached to the bottom of the rear outer interplane strut, fluttering forward in the wind.

Facing Page, Below: Three Halberstadt D III are jacked up on the airfield Roupy. In the foreground, at far left, the Halberstadt D III of Oblt. von Althaus can be seen. On the right of the photo, an officer scans the sky for enemy aircraft. These two photos were likely taken in late November or December 1916.

Lt. Ernst von Althaus, Halberstadt D III, autumn 1916, Profile 36

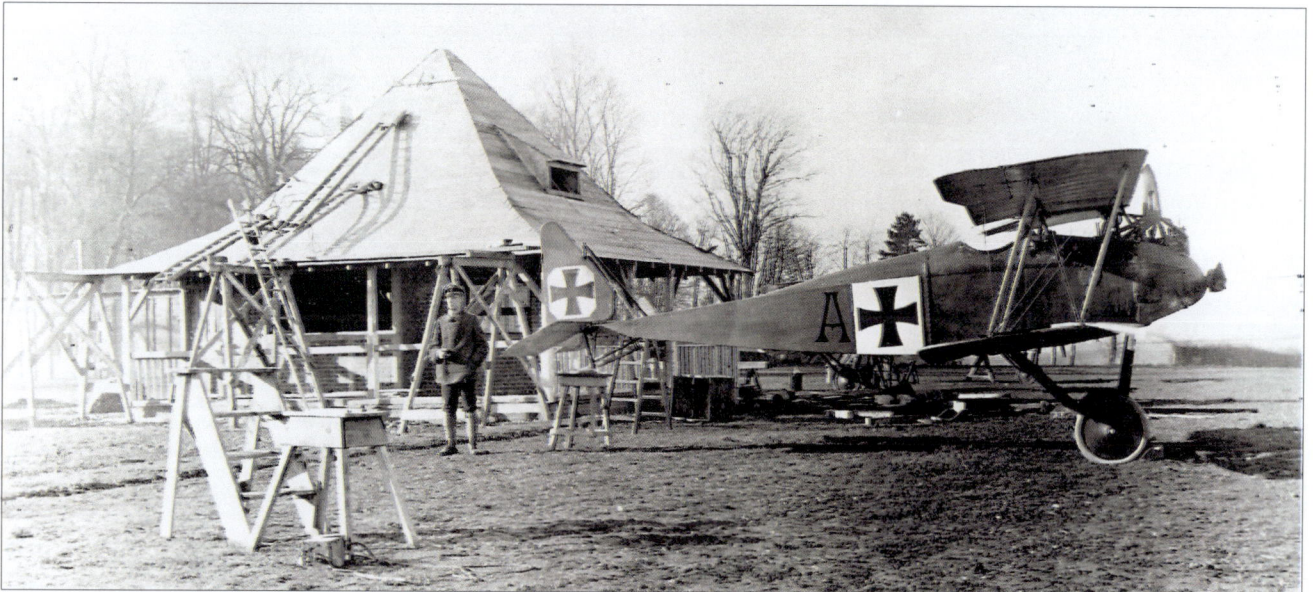

he had already achieved two aerial victories by the end of the year. On 20 March 1916, on the occasion of his fourth aerial victory, he was mentioned for the first time in the German Army Report. In April 1916, he was transferred to Fokkerstaffel Jametz at the Verdun front, but after two more aerial victories, he returned to his old unit in the Somme area in mid-June. In the meantime, he had also been awarded the Knight's Cross of the Royal House Order of Hohenzollern.

After his 8th aerial victory on 21 July 1916, he became the fifth aviator to be awarded the Order Pour le Mérite. When Jagdstaffel 4 was formed from Kampfeinsitzer-Kommando Vaux on 25 August 1916, he was part of the founding cadre of this new unit. In October 1916 he met his future wife, the nurse Emmy Schulz. From 11 November 1916, he served as an instructor at the Jagdstaffelschule I in Valenciennes for about three months.

Back with Jagdstaffel 4, he was wounded by a shot in the foot in aerial combat on 4 March 1917 and had to spend several weeks in a military hospital. After his recovery, he joined his old comrade in arms Rudolf Berthold's Jagdstaffel 14 as deputy Staffelführer on 25 April. On 6 July 1917, at Manfred von Richthofen's request, he took over as Staffelführer of Jagdstaffel 10 in his Jagdgeschwader I, where he achieved his ninth and last aerial victory on 24 July. However, he had to relinquish command of the Staffel on 30 July 1917 and was transferred to the Jagdstaffelschule II at Nivelles as commander.[7]

Fighter pilot Alfred Lenz, who served alongside Ernst von Althaus in Jagdstaffel 4, described him to Alex Imrie as a person of charm, elegance, and winning demeanor. At the same time, however, he called him: *"the most light-headed person I have ever met"*.

This character trait was, according to Lenz: *"coupled with an almost childlike credulity and naivety. In addition, he was a passionate gambler who was always outplayed and, accordingly, was constantly in need of money."*[8]

It was probably for this reason that he got involved in illegal, dubious businesses. These illegal dealings were uncovered, and in December 1917 Oberleutnant of the Reserve Freiherr von Althaus was sentenced by the field war court of the stage inspection 2 to one year and three months of imprisonment for dealing with embezzled army property. In this situation, he was helped both by the fact that he was a Knight of the Order Pour le Mérite and by his father's name and position. The King of Saxony soon pardoned the overthrown hero.[9]

Above: Oblt. Ernst Freiherr von Althaus in the uniform of the Saxon Hussar-Regiment „König Albert" Nr. 18.

Ernst von Althaus fought again in the infantry from August 1918 as a platoon and company commander, where he showed great courage and steadfastness. On 15 October 1918, he fell into American captivity. After his release from captivity in September 1919, he studied law, obtained his doctorate, and practiced law. Due to the new republican legal situation, he was able to take on the name of his father and from then on lived under the name Ernst Prinz zu Bentheim und Steinfurt. He continued to practice his profession as a lawyer after becoming affected by total blindness in 1937. During World War II he was – in spite of his disability – president of the court in Berlin. From the summer of 1945 he is said to have worked as a translator for the Allies, but this is not documented. He died in Berlin on 29 November 1946.[10]

According to available records, most Halberstadt and Albatros D IIs of Jagdstaffel 4 in the period August 1916 – Winter 1917 initially had no personal insignia. According to the available photos, personal identifications only began to appear in early 1917 when the Staffel received the Albatros D III.[11]

7.5 Royal Prussian Jagdstaffel 5

Staffelführer:

Oblt. Hans Berr 21 August 1916 – 06 April 1917[1]

Jagdstaffel 5 is one of the best documented single-seater units of the German Fliegertruppe. One of the first fighter pilots I was able to interview several times was the later rocket pioneer Rudolf Nebel, whom I interviewed in 1976 and 1977. He had been transferred to Jagdstaffel 5 on 30 November 1916, which at that time was stationed at Gonnelieu Airfield in the German 1st Army area. He belonged to the Staffel until 31 July 1917.[2]

To my great pleasure he possessed an extensive photo album, which documented the history of the unit from November 1916 to July 1917 very well. Furthermore, Herr Nebel still had his flight log, as well as extensive written documentation about Jagdstaffel 5.

The Staffel equipment consisted of the Halberstadt D III, which, according to Rudolf Nebel, was a sensitive aircraft that flew well:

"When I arrived at the Staffel it was uniformly equipped with the Halberstadt. I was assigned to Geschwader I, Gruppe 3. My Gruppenführer was Lt. Renatus Theiller, a friendly and humorous man from Alsace, at the same time an excellent fighter pilot and instructor.

Above: Lt. Rudolf Nebel in the pilot's seat of a Roland D II at the Jagdstaffel School Valenciennes in January 1917. His recollections and documents of Jagdstaffel 5 are invaluable to the history of this Staffel.

In the Staffel there were precise instructions on how to fly as a group or Geschwader. Hans Berr

Above: Pilots of Jagdstaffel 5 at Gonnelieu airfield in November 1916. Seen third from left is Vizeflugmaat Winkelmann, fourth from left Vzfw. Büssing. The blue-gray Halberstadt D III carrying the white number "1" seen in background was the personal mount of Oblt. Hans Berr. On the right is the Halberstadt D III marked with the number "3".

**Oblt. Hans Berr, Halberstadt D III
Jagdstaffel 5, November 1916, Profile 37**

attached great importance to our training. After each flight to the front, there was a briefing. Berr also criticized us, but always in a benevolent, constructive manner. Berr was a very friendly and polite man whom we trusted completely. He was my role model when I finally got my own Staffeln (KeSt. 1b, KeSt. 1a and Jagdstaffel 90, author's note)."[3]

Oblt. Hans Berr, Halberstadt D III Jagdstaffel 5, November 1916, Profile 37

It had a white."1" as Staffelführer on both sides of the fuselage and on the top of the fuselage. Some of the Halberstadt D IIIs that served with Jagdstaffel 5 had a fabric covering that appears medium-dark gray in the photos. According to Rudolf Nebel, this fabric covering was a darker blue gray.[4]

Hans Berr was born on 20 May 1890 in Magdeburg, in the province of Saxony in the Kingdom of Prussia. He started his career as a soldier before the war and was promoted to officer in 1908 with the 4th Magdeburg Infanterie Regiment No. 67. At the outbreak of war, he went into the field with the 7th Rhenish Infanterie Regiment No. 69 and was severely wounded on 6 September. Promoted to Oberleutnant on 27 January 1915, he joined the Fliegertruppe in March of that year. Trained as an observer at Flieger-Ersatz-Abteilung 6 in Großenhain, he was on frontline duty with Artillerieflieger-Abteilung 211 in Lorraine from May 1915 on. In the fall of 1915, he was trained as an pilot at the Metz Flying School, then briefly returned to the front and at the end of the year took over as leader Kampfeinsitzer-Schule in Cambrai. With the formation of the first Kampfeinsitzer-Kommandos and the

Fokkerstaffeln at the beginning of 1916, however, he returned to frontline operations and was initially leader of Fokkerstaffel Jametz in the front section before Verdun, where he achieved two aerial victories in a short time. In the middle of the year, he took over the Fokkerstaffel Avillers in the same front sector.

On 21 August 1916, Jagdstaffel 5 was formed from the flying personnel of Fokkerstaffel Avillers and Hans Berr became Staffelführer of this unit. At the end of September the Staffel moved to the Somme and here he was very successful with another eight aerial victories in only four weeks. On 10 November 1916, he was then awarded the Knight's Cross of the Royal House Order of Hohenzollern; on 4 December, he was awarded the Order Pour le Mérite.

On 6 April 1917, while flying at the head of his Staffel, he was rammed by Vzfw. Paul Hoppe while attacking an F.E. unit, who was about to attack the same aircraft, and crashed fatally at the controls of Albatros D III D.2106/16 near Noyelles. Vzfw. Hoppe also met his death. In addition to the awards already mentioned, Berr also received the Bavarian Military Order of Merit IV Class with swords, the Prussian War Merit and Honor Cross 1st Class, the Brunswick War Merit Cross II Class and the Hamburg Hanseatic Cross during his military career.[5]

Vzfw. Hans Müller, Halberstadt D III, Jagdstaffel 5, November 1916, Profile 38

The Halberstadt D III of the leader of Gruppe 4 had a light blue fabric covering like the machine of Rudolf Nebel. The letter "C" was red like all numbers and letters on the light blue Halberstadt fighters flown by Jagdstaffel 5.[6]

Right: Oblt. Hans Berr seated in his office at the Villa Boistrancourt. Jagdstaffel 5 was based at Boistrancourt for just over a year from early March 1917 to late March 1918.

Vzfw. Hans Müller, Halberstadt D III, Jagdstaffel 5, November 1916, Profile 38

Hans Müller was born on 19 July 1892, in Loschwitz, now a district of Dresden, in the Kingdom of Saxony. He enlisted in the Fliegertruppe as a war volunteer at the outbreak of war and was trained as an airplane pilot at Flieger-Ersatz-Abteilung 2 in Schneidemühl until the end of the year. He was transferred on 3 January 1915 as a Unteroffizier to Feldflieger-Abteilung 3 in the Arras battle area, where he won one of the first aerial victories of the German Fliegertruppe on 26 March 1915. On 1 May, via Army Flight Park 6, he joined Flieger-Ersatz-Abteilung 6 at Großenhain as an instructor.

Promoted in the meantime to Vizefeldwebel on 20 February 1916, he was transferred to Kampfstaffel 11 in Kagohl 2 on the Verdun front. In April 1916 he retrained as a fighter pilot and joined the Fokkerstaffel Avillers on 28 June. In the two months that followed, he scored two aerial victories.

After the formation of Jagdstaffel 5 on 21 August 1916, he scored what was perhaps the first aerial victory of the new Staffel on 31 August. In September, Jagdstaffel 5 moved to the Somme and by the end of the year he had achieved five more aerial victories, making him one of the Staffel's most successful pilots. Badly wounded in air combat on 26 December 1916, by a bullet in the stomach,

Above: The sky-blue Halberstadt D III flown by Vzfw. Hans Müller. It was marked with a red "C" as personal marking. In the background are two other sky-blue Halberstadt D IIIs.

Right: Sanke card No. 446, "Our successful fighter pilot Vzfw. Hans Müller", seen here following his promotion to the rank of Leutnant. Between 26 August and 26 December 1916, Müller was credited with a total of 9 victories.

he received word of his promotion to Leutnant in hospital on 14 January 1917.

No longer in frontline service after his recovery, he joined Siemens-Schuckert Flugzeugwerke as an acceptance and demonstration pilot in mid-1917. During his frontline flying career, both the E.K. II as well as the E.K. I were awarded to him. After the war, he emigrated to Mexico, where he operated a flying school until 1931. He then settled down in Texas. He died in San Antonio, Texas, on 23 July 1977.[7]

It should be noted that Jagdstaffel 5 was never equipped with Roland D II aircraft. The Roland D IIs assigned to Jagdstaffel 5 in various publications are photos of unarmed training aircraft serving with Jagdstaffelschule Valenciennes, originating from the photo album of Rudolf Nebel, as **he personally confirmed to me.**

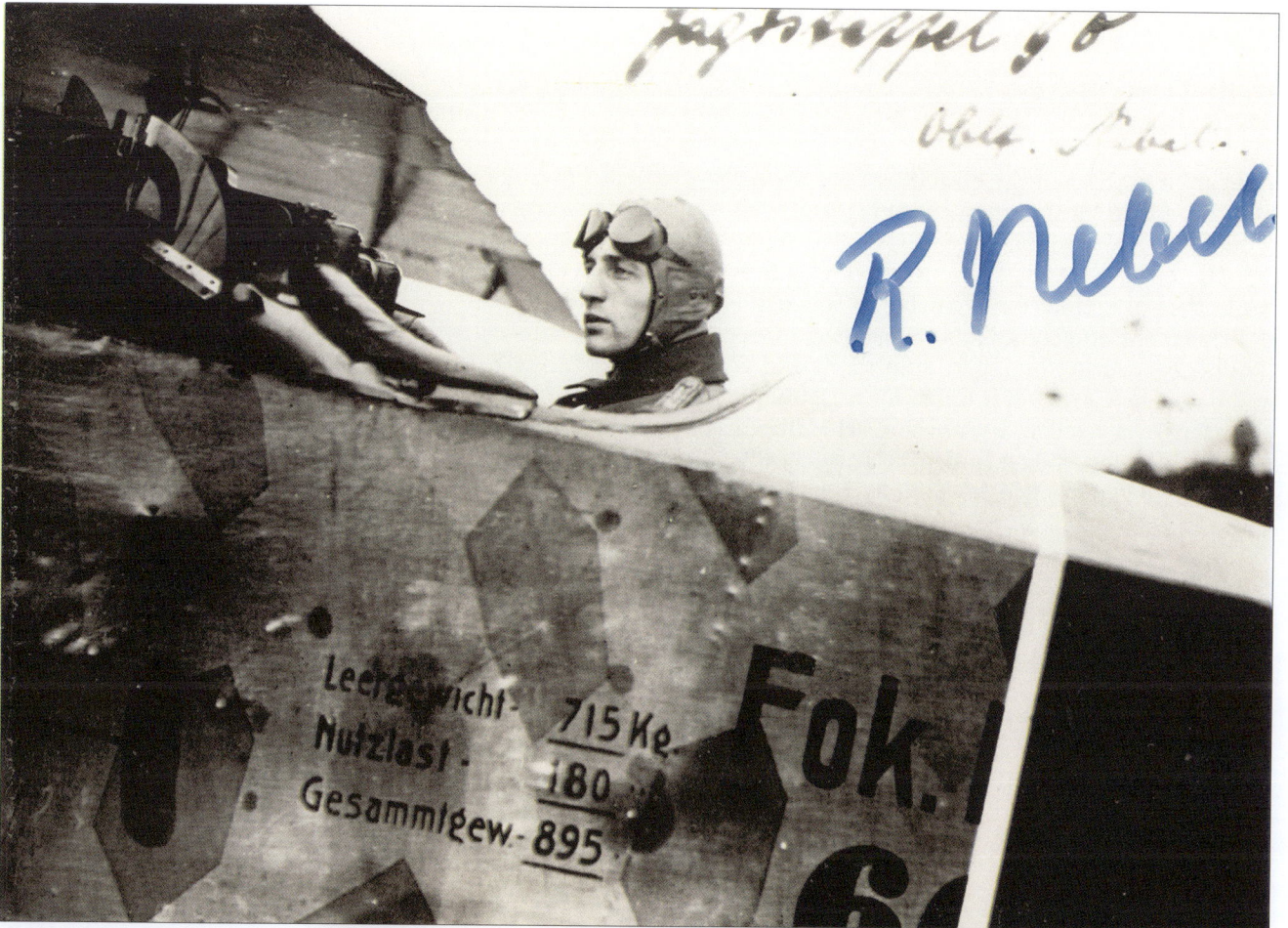

Above: Signed photo from Rudolf Nebel for Bruno Schmäling.

Right: Rudolf Nebel and Bruno Schmäling.

7.6 Royal Prussian Jagdstaffel 6

Staffelführer:
Rittm. Josef Wulff   25 August 1916 – 1 May 1917[1]

The color profiles of the aircraft are based on the photos of Carl Holler, member of Jagdstaffel 6, which my friend and colleague Greg VanWyngarden kindly made available to me, as well as photos from the collection of my colleague Tobias Weber. Furthermore, the book "*Als Sängerflieger im Weltkrieg*" (As A Singing Pilot in the World War) published by Carl Holler under his artist name "Nils Sörnsen" was used as a source.[2] Carl Holler was a well-known folk singer under the stage name "Nils Sörnsen". Using today's terminology, he would be called a "country star" in the U.S.A., which is why he was given the name "Sänger-Flieger" (Singing Pilot) by his fellow airmen.

The unit had arrived at Ugny l' Equipée airfield in the German 2nd Army area on 30 September.[3] According to Carl Holler, the Staffel received the new Albatros D I at this airfield. The available photos and documents show that each aircraft received a number as individual identification, although it is difficult to determine at what time these markings were first applied. Rittm. Wulff ordered a system of markings for the aircraft consisting of the numbers 1 to 12. In December 1916 the unit began to receive Albatros D II fighters. According to the photos from Carl Holler the Staffel got some of the Albatros D IIs produced by the Ostdeutsche Albatros Werke.

Vzfw. Carl Holler, Albatros D II (O.A.W.), December 1916 - January 1917, Profile 39 & 39a

The aircraft is one of the 50 Albatros D IIs built at the Ostdeutsche Albatros Werke (O.A.W.) in the military numbers range 890/16 - 939/16.[4] These can be identified by the position of the metal plate on the side of the pilot's seat with the production data, as well as the fuselage Iron Cross being positioned further forward compared to Albatros- and L.V.G.-built examples of the type.[5] The aircraft has a factory-applied camouflage paint scheme. This can be seen by the soft color transitions, which may have been light green and dark green or light green, dark green, and rust red. In the profile, we opted for a light green, dark green, and rust red camouflage. The personal identification "9" was painted in black on a white rectangle.

Above: Pilots of Jagdstaffel 6 in front of Albatros D I D.450/16, from left: 1. Vzfw. Petzold, 2. (unknown), 3. Oblt. Volkmann (Offz.z.b.V.), 4. Rittm. Wulff (Staffelführer), 5. Oblt. Reinhold, 6. (unknown) 7. Vzfw. Fritz Loerzer, 8. (unknown). Seated on the cockpit padding is Vzfw. Carl Holler.

Above: The Albatros D II (O.A.W.) of Vzfw. Carl Holler in January/February 1917 with the factory applied camouflage and the black number "9" on a white rectangular field as personal marking. (G. VanWyngarden)

Carl Holler was born on 26 July 1884, in Rendsburg in Holstein in the north of Germany. As early as the fall of 1914, he enlisted in the Fliegertruppe and was transferred to the Flieger-Ersatz-Abteilung 1 in Döberitz in November 1914 for training as a pilot. He received his flight training at the Fliegerschule Niederneudorf. After his promotion to Vizefeldwebel, he was transferred to Armee-Flug-Park 13 in Serbia on 16 September 1915, and from there to Feldflieger-Abteilung 1 in January 1916. There he also received the Iron Cross 2nd Class. In the spring of 1916, he reported for training as a single-seat fighter pilot and in May 1916 was assigned to the Kampfeinsitzer-Schule Köln. After a short training period, he first joined Fokkerstaffel Jametz in the German 5th Army area. Barely two weeks later, he was transferred to Fokkerstaffel Sivry. When Jagdstaffel 6 was formed, he was transferred to this unit, as were other pilots from Fokkerstaffel Sivry. The official date of his transfer is 28 August 1916. On 25 January 1917, Carl Holler

Left: Vzfw. Carl Holler, aka Nils Sörnsen. (G. VanWyngarden)

Vzfw. Carl Holler, Albatros D II (O.A.W.), December 1916 - January 1917, Profile 39 & 39a

reported shooting down a Nieuport, southwest of Roye, but it was only recognized as "forced to land beyond the lines." The same fate befell him on 11 February 1917, when the Caudron he shot down was again recognized only as "forced to land beyond the lines".

On 1 May 1917, the leadership of the Staffel changed from Rittm. Wulff to Lt. Otto Bernert. Under the leadership of his new Staffelführer, Carl Holler achieved his first and second aerial victories on 19 May 1917 and 30 May 1917.[6] In his book "Als Sänger-Flieger im Weltkrieg," Carl Holler wrote that after one of the aerial victories he got into trouble with an officer pilot who disputed his victory. After checking the facts, however, the Staffelführer Lt. Bernert awarded the aerial victory to Carl Holler. After Otto Bernert fell ill, this officer pilot took advantage of the absence of Otto Bernert, who was considered fair and just, and harassed Carl Holler

to such an extent that he asked to be transferred out of the Staffel. He was sent to Flieger-Ersatz-Abteilung 11 in Breslau where he later met his friend and former Staffelführer Otto Bernert again. On 17 October 1917, he was transferred to the 2. Seeflieger-Abteilung (naval-flyer) at Wilhelmshaven.[7]

After the war he devoted himself again to his singing career, he founded a music publishing house and was active in the German Aviation Sports Association. He passed away in Hamburg on 16 December 1966.

The records of Carl Holler, published as a book, vividly show that the activities of enemy airmen were not the only threats he and his fellow pilots had to face. Problems brought about by the new technologies built into their own aircraft could also lead to dangerous situations. Thus reported Vzfw. Holler on 15 November 1916:

"I took off this morning at 8:06 as the third plane

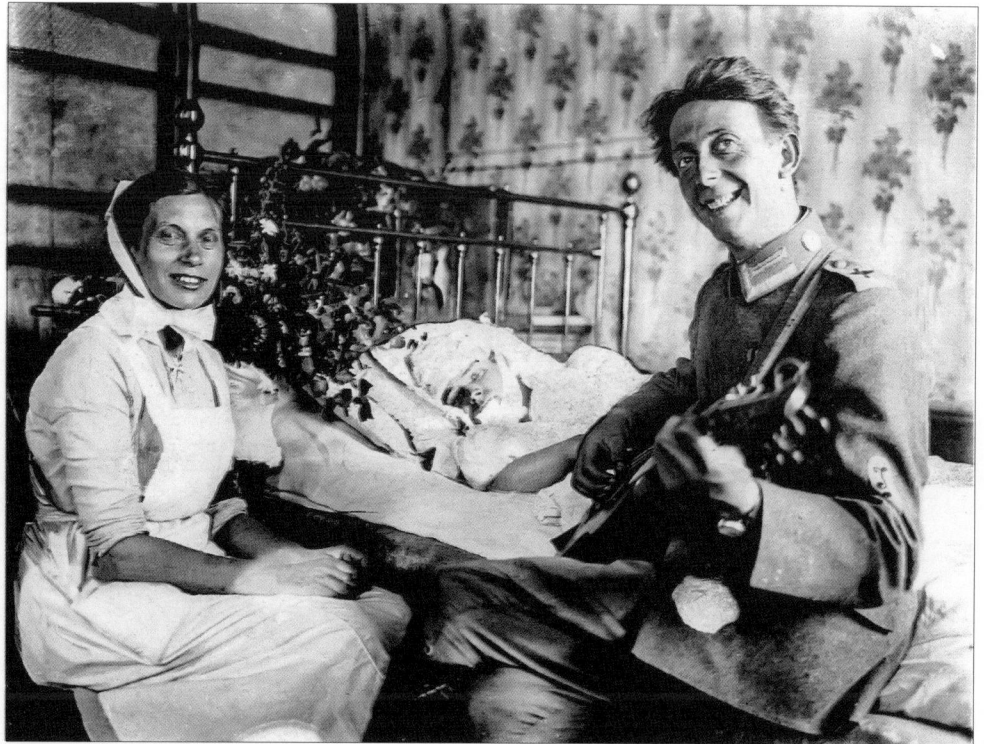

Right: Carl Holler sings and plays the mandolin by the side of Otto Bernert's sickbed. The latter was in the process of recovering from severe injuries he had sustained as the result of a crash on 21 May 1917. (G. VanWyngarden)

of the first Kette, which was led by Lt. Fr. (Lt. Bruno Freitag, author's note) and had the task of blocking the front from enemy reconnaissance planes at the highest possible altitude during the simultaneous attack on Pressoir Forest. Immediately after the approach we came over a closed, very low cloud cover, I lost orientation and followed the leader aircraft (the third had already separated). After the end of the barrier time [intended duration of the blocking flight at the front, author's note] the leader of the Kette gave me the signal to fly home, whereupon we changed from 4,300 meters to gliding flight. After a short time, it was impossible for me to see through my glasses because they were covered by oil, and the leader plane disappeared from my sight. I removed the goggles. Now the oil was splashing into my eyes, which made flying even more difficult. I pushed through the very low cloud cover, which reached down to 60 meters above the ground.

Near a village I didn't know, which I circled three times, I chose a meadow as an emergency landing site. I taxied up to a tree-lined road. When the plane had come to a halt, I switched on late ignition, turned off the engine and took off my fur boots. The oil feed tube to the camshaft was broken, so the pump just managed to splatter the oil onto my face. I now re-routed the gasoline supply from the emergency tank into the main tank, then disconnected a hose connection between the two and used it to patch the broken oil pipe. While I was removing the oil from my glasses, I recognized the soldiers approaching me from several sides as Frenchmen. The situation was immediately clear to me. I was behind enemy lines. Fortunately, my engine had held compression, and since I could not take off in the direction of the trees, I first had to make a U-turn on the ground and took off in the opposite direction under heavy rifle fire, which did not stop until I was back in the clouds. Now I flew exactly in an eastern direction, but after 25 minutes I had to make a second emergency landing, because the hose connection had loosened again, and the engine started to spew oil again. However, I could just see Russian prisoners of war guarded by German soldiers working in the field from the air. I was near Hamm on the Somme, took off again, but the area of my airport was covered with dense ground fog, which forced me to make a third emergency landing near Ennemain. Several more German planes landed there after me for the same reason. I reached our airport at 11:40.[8]

7.7 Royal Prussian Jagdstaffel 7

Staffelführer:

Oblt. Fritz von Bronsart and Schellendorf 22 August 1916 – 21 July 1917[1]

The presentation of the markings of the aircraft are based on the photos in the photo albums of the former Staffel members Lt. Wilhelm Eckenberg, Oberflugmeister Kurt Schönfelder and Vzfw. Friedrich Mannschott.

In the 1980's I received the address of Dr. Edith Freifrau von und zu der Weichs, the daughter of Lt. Wilhelm Eckenberg, former member of the Artillerieflieger-Abteilung 203, Kampfeinsitzer-Kommando Avillers, as well as Jagdstaffeln 7, 32, and 29. Shortly afterwards I was able to visit Frau Dr. Edith von und zu der Weichs at Schloss Weichs. She kindly provided me with a beautiful photo album of her father with interesting photos of the above-mentioned units.

A few years earlier, after a long search, I had located relatives of Friedrich Mannschott, the most successful pilot from the early days of Jagdstaffel 7. The Engelhardt family was in possession of a very nice photo album with photos of Artillerieflieger-Abteilung 203 and Jagdstaffel 7, which I was able to copy.

A particularly valuable find was the small photo album of Oberflugmeister Kurt Schönfelder, Jagdstaffel 7, which I was also able to photograph. In the photo album there were also three photos from the early days of Jagdstaffel 7, from the time before he himself joined Jagdstaffel 7.

According to these photos and documents, Jagdstaffel 7 was initially equipped with Fokker E IIIs and E IVs, which were later replaced by Albatros D IIs.[2]

Lt. Wilhelm Eckenberg, Albatros D II D.1793/16, late autumn 1916, Profile 40

A photo in the photo album of Wilhelm Eckenberg shows him in the pilot's seat of an Albatros D II at Jagdstaffel 7, with a "pipe-smoking moon" as his personal insignia. The fuselage of the Albatros was left in the usual varnished "honey-colored" plywood, and the color of the moon was interpreted as an "ivory-white" color.

Wilhelm Eckenberg was born on 13 November 1895 in Hüllen near Gelsenkirchen, which actually became a part of the city of Gelsenkirchen eight years later. After enlisting in the Fliegertruppe, his training as a pilot began on 24 April 1915, with

Above: Vzfw. Friedrich Mannschott was the first successful fighter pilot of Jagdstaffel 7, being credited with 12 victories between 05 January and 16 March 1917. He was killed during a balloon-attack on the latter date.

Flieger-Ersatz-Abteilung 5. From 10 May 1915, his further training took place at the Militärflieger-Schule Hamburg-Fuhlsbüttel, from where he returned to Flieger-Ersatz-Abteilung 5 on 28 July 1915. After his promotion to Unteroffizier on 18 August 1915, he was transferred to Artillerieflieger-Abteilung 203 on 6 September 1915, where he earned his pilots badge on 20 October. On 27 January 1916, he was promoted to Vizefeldwebel and on 20 July 1916, to Leutnant. On 9 July 1916, he was wounded in aerial combat as single-seat fighter pilot

Above: Albatros D II D.1793/16 was one of the last aircraft of this variant to be produced by the parent company, with D.1799/16 being the ultimate example. It became the personal plane of Lt. Wilhelm Eckenberg's Albatros D II at Jagdstaffel 7, who chose a "smoke piping man in the moon" as his personal marking. The muddy conditions of the airfield in early 1917, most likely near Spincourt, are apparent.

Lt. Wilhelm Eckenberg, Albatros D II D.1793/16, late autumn 1916, Profile 40

and was awarded the Iron Cross I Class on 10 August 1916, for his service.

His transfer to Jagdstaffel 7 took place on 9 October 1916, to which he belonged for a good ten months. On 14 July 1917 he was transferred to Jagdstaffel 32 and on 7 September 1917 to Jagdstaffel 29.[3] How long he remained a member of Jagdstaffel 29 is not clear from the available transcript of the war diary of Jagdstaffel 29.

Unfortunately, there is no more information on the markings of the Jagdstaffel in the year 1916. At some time in early 1917 the Staffel took delivery of the first Albatros D III.

Right: Lt. Wilhelm Eckenberg.

7.8 Royal Prussian Jagdstaffel 8

Staffelführer:

Hptm. Gustav Stenzel   23 September 1916 – 28 July 1917[1]

Jagdstaffel 8 is also one of the poorly documented early Jagdstaffeln of the German Fliegertruppe. Except for Erich Tornuß' transcript of its war diary, hardly any documents are available. For a long time, only four photos of the Staffel were available, which Dr. Bock had received from the family of former Staffel member Alfred Ulmer, with whom he was in contact in the 1970s. Thus, this chapter about the early days of Jagdstaffel 8 would have remained without any color profile, had not Greg VanWyngarden generously made his research and the corresponding photos available to me!

Lt. Walter Goettsch, Albatros D II, Autumn 1916, Profile 41

The aircraft of Jagdstaffel 8 were divided into "Ketten" in the fall of 1916. Lt. Walter Goettsch had ordered that all aircraft in his Ketten should have the fuselage painted white. Accordingly, the fuselage of his Albatros D II was also white. The photo gives the impression that it was a low-quality or thinly applied paint, since the wooden structure of the fuselage still shows through.

This is the earliest documented case of the aircraft in a Ketten or group having a uniform color as marking.[2]

Walter Goettsch was born in Altona near Hamburg on 10 June 1896. After graduating from high school, he enlisted as a war volunteer in July 1915 and was trained as a pilot at Flieger-Ersatz-Abteilung 7 in Cologne and the Fliegerschule Mannheim. On 4 April 1916, he joined Feldflieger-Abteilung 33 in Flanders as a Unteroffizier, quickly earning the E.K. II and the E.K. I and retrained there in the summer for service on single-seater fighter airplanes.

On 12 September 1916, he was transferred to Jagdstaffel 8 as the unit was initially formed. Promoted to Vizefeldwebel the following month, he achieved his first two aerial victories in November. In January 1917, he was appointed Offizierstellvertreter and, after six aerial victories, was slightly wounded in aerial combat over Wervicq on 3 February 1917. Promoted to Leutnant for bravery in the face of the enemy after 12 aerial victories on 2 June 1917, he was awarded the Knight's Cross of the Royal House Order of Hohenzollern after his 14th kill on 23 August 1917.

Above: The first Staffelführer of Jagdstaffel 8, Hptm. Gustav Stenzel.

In September he scored three more aerial victories and on 12 October 1917, was placed at the disposal of the Inspektion der Flieger and transferred to Flieber-Ersatz-Abteilung 4 as an instructor.

Right: Vzfw. Walter Göttsch in front of his white painted Albatros D II. The quality of the paint does not seem to have been very good or it was thinly applied, as the plywood structure below is still visible. A retouched version of this photo appeared as "Sanke" card No. 518. (G. VanWyngarden)

On 14 February 1918, he was appointed Staffelführer of Jagdstaffel 19 in Jagdgeschwader II. During the German spring offensive, he achieved two aerial victories. On 10 April 1918, while on a protective flight for bombers attacking Amiens, after a victorious aerial combat with an R.E. 8, he "stalled" on Fokker Dr. I 419/17 in a right turn from low altitude due to excessive loss of speed and crashed fatally near the Genteiles forest. His plane caught fire on impact and burned completely.[3]

Unfortunately, also for the later period, spring 1917 until the end of the war, almost no photographic material of Jagdstaffel 8 is available. This is all the more regrettable as I had the opportunity to meet and interview the last Staffelfürer of Jagdstaffel 8, General (ret.) Werner Junck, several times in Munich in 1974. He had been transferred to Jagdstaffel 8 on 12 February 1917 and belonged to the Staffel until the end of the war. From 1 April 1918, he was Staffelführer.

I had found his address and telephone number in the membership directory of the "Association of Old Eagles", of which he was chairman for many years. When I contacted him by telephone, to my surprise he arranged to meet me in his favorite pub, where he liked to have his "nightcap", as he called it, in the evening.

When I introduced myself to him, he greeted me because of my long hair, in keeping with the fashion of the time, and colorful tie-dye T-shirt, saying, *"I didn't think a hippie would be interested in us old flyers."*

I was a bit perplexed at first, then he grinned, shook my hand, and invited me to take a seat with him. No sooner had I sat down than I was invited for a beer. In the conversation he had to tell me to his great regret that his extensive records and his photo albums from the time of Jagdstaffel 8 had been

Lt. Walter Goettsch, Albatros D II, Autumn 1916, Profile 41

Above: Lt. Alfred Traeger seated in the cockpit of his Albatros D II. The white fuselage seen on his aircraft was the distinctive marking of "Kette Göttsch". (G. VanWyngarden)

Above: Vzfw. Walter Göttsch joined Jagdstaffel 8 on 15 September 1916, and Fokker D II 229/16 was apparently the first fighter-aircraft assigned to him at his new unit. (G. VanWyngarden)

Above: NCO-pilots of Jagdstaffel 8 at Rumbeke airfield, from left: 1. Uffz. Seitz, 2. Uffz. Felming, 3. Uffz. Goettsch, 4. Vzfw. Weichel, 5. Uffz. Hermann. In the background a newly delivered Albatros D I can be seen. This was another prototype Albatros D I fitted with a water header tank in front of the first engine cylinder. (G. VanWyngarden)

completely destroyed in a bombing raid on Berlin in World War II. This was a loss that still hurt him very much 30 years later. Several times in conversation he shook his head and said:

"How gladly I would have made all my documents available to you. I think it is great that such a young man in Germany is interested in our history. How nice it would have been if you had written the history of my Jagdstaffel."

He shared a lot of information with me concerning his Jagdstaffel 8 and the painting schemes applied to the aircraft. In this context I unfortunately have to state that no previously published color drawing of a Jagdstaffel 8 aircraft from the period of time when the Staffel was under his command matches this information!

I met with him a few more times for "a beer", as he called it. For this reason, I regret that no photos of Jagdstaffel 8 have ever surfaced. If photos do turn up, I could contribute a sufficient amount of information concerning painting schemes of the aircraft, which came from my friendly interlocutor.

7.9 Royal Prussian Jagdstaffel 9

Staffelführer:

Oblt. Kurt Student 05 October 1916 – 14 March 1918[1]

The history of Jagdstaffel 9 is very well documented. The membership directory of the "Gemeinschaft der Alten Adler" also contained the address of retired Gen. Kurt Student, who lived in Bad Salzuflen. Accordingly, he was one of the first fighter pilots I visited and interviewed. Although he was busy working on his book about the German World War II paratroopers, which he had been instrumental in building, he took time for a lengthy conversation. He also had an extensive photo album, which I was allowed to copy. The notes of our conversation are still available, unfortunately the notes of the photo captions of the photo album have been partially lost in the course of time. If someone could help me out with this, I would be very grateful.

Further documents and photos from the early days of the Staffel come from the photo album of the fighter pilot Hermann Pfeiffer, which Michael Schmeelke received from the relatives of the pilot and was able to copy, as well as the photo album of Hartmut Baldamus whose photos were kindly made available to me by Hannes Täger. Herr Gen. Student also gave me the address of Hans Mohr, one of his mechanics at Jagdstaffel 9. Fortunately, his son could still be found at the same address, and so I was able to copy his photos.

According to the transcript of the war diary, Jagdstaffel 9 initially took over four Fokker E I and seven Fokker E III aircraft from the "Armeestaffel des A.O.K. (Armee-Ober-Kommando der deutschen 3. Armee)". In addition, the Staffel had a Fokker D II, a Fokker D III and a captured Nieuport 17 at its disposal. A Rumpler C I was available as well, serving as the Staffel "hack" aircraft.[2] The Fokker E I and E III were soon replaced by additional Fokker D II and Halberstadt D II/ DIII aircraft. In early 1917, the Staffel received Albatros D IIs, and many of these were of the variant built under license by L.V.G.

As a photo and the corresponding note in Hans Mohr's album show, the Staffel had three new Albatros D IIIs in February 1917, in addition to the Albatros D II.[3]

According to Gen. (ret.) Student, the Albatros D IIs performed considerably better than the Fokker D IIs, Halberstadt D IIs and D IIIs. Accordingly, the airmen were enthusiastic about their new Albatros D IIs. This is also shown by the fact that the pilots quickly applied personal markings and paint schemes to their Albatros D IIs, whereas no evidence

Above: Vzfw. Erich Köhler in front of his Fokker D II in September 1916.

of these can be found on the previous Fokker and Halberstadt fighter planes.

Oblt. Kurt Student, Albatros D II (L.V.G.), February 1917, Profile 42

On the fuselage of his first Albatros D II (L.V.G.) he carried a badge, which I first interpreted as an "observer badge". This surprised me since Kurt Student had never been an observer. In our conversation he corrected me that the badge stood for the "Armeestaffel des A.O.K. 3" and represented the military sign for an Army High Command:[4]

"We (Armeestaffel des A.O.K. 3, author's note) were directly subordinated to the A.O.K. and were very well supported. Frequently I attended meetings at A.O.K. The good cooperation, also concerning the Flugmeldedienst (flight reporting service), was a reason for the success of the Staffel. Even as Jasta 9, we were still the A.O.K. Staffel."[5]

This aircraft was one of 75 Albatros D II fighters that were manufactured **under license by L.V.G.. The fuselages of the L.V.G.-built examples of this fighter that served with Jagdstaffel 9 all appear comparatively dark in the many available photos, and also in the motion picture footage that was filmed at the unit's airfield at Leffincourt by the**

Above: Lt. Schlolaut together with his mechanics in front of his newly delivered Halberstadt D II.

**Oblt. Kurt Student, Albatros D II
(L.V.G.), February 1917, Profile 42**

Above: Jagdstaffel 9 received several LVG-built examples of the Albatros D II. Here a trio of these are warming-up their engines a the airfield at Leffincourt in early 1917. It appears that the fuselages of these aircraft are rust-red. The first machine in the line-up is that of Oblt. Student, the second was flown by Vzfw. Köhler.

Bufa (Bild-und Film-Amt). Since there are numerous reports of former Jagdstaffel pilots that some Albatros D II fighters had a "rust red" or "reddish brown" fuselage, it appears that these L.V.G.-built D IIs were delivered with plywood schemes that had been coated with a dark-tinted wood stain. The color profiles have been prepared accordingly.

Kurt Student was born on 12 May 1890, in Birkholz/Kreis Züllichau-Schwiebus in the Prussian province of Brandenburg (now Lebus Voivodeship, Poland). He joined the Jäger Battalion "Graf Yorck von Wartenburg" No. 1 as an officer candidate after graduating from high school on 3 March 1910 and completed the officer training course at the Kriegsschule Danzig from May 1910 to March 1911. Promoted to Leutnant on 20 March 1911, he was a Kompanieoffizier in his battalion for the next two years. On 1 August 1913, he reported for training as a pilot with the Fliegertruppe, which was being formed, and received his training until January 1914 at the Militär-Fliegerschule Johannisthal. From February 1914 he served as a pilot with Flieger Battalion 2 in Posen, he was transferred to Feldflieger-Abteilung 17 at the outbreak of war and fought with it for the next year and a half, mainly in Russia, soon receiving the E.K. II and the E.K. I.

On 15 June 1915, he was awarded the Knight's Cross 2nd Class with Swords of the Saxon Order of Albert and three days later was promoted to Oberleutnant. Three months later, on 30 September,

he managed to shoot down a Russian Morane in Galicia, but he was not credited with this kill. Transferred to France in early 1916, he flew with Kagohl IV's Kampfstaffel 19 at Verdun from 10 February and was retrained as a fighter pilot in May. On 1 June 1916, he became leader of Fokker-Staffel A in Champagne, which was renamed Armeestaffel des A.O.K. 3 the following month. This Staffel achieved great success over the next four months, including his own initial two aerial victories.

When the first Jagdstaffeln were formed in the fall of 1916, the Fokkerstaffel was renamed Jagdstaffel 9 on 28 September, and he was officially appointed Staffelführer of the "new" Staffel on 7 October. Lightly wounded twice after his 3rd aerial victory (first in a crash with an S.S.W. D I on 15 April 1917, and three weeks later, on 2 May 1917, in aerial combat), he was awarded the Knight's Cross of the Royal House Order of Hohenzollern with swords on 5 June 1917, in recognition of his achievements as Staffelführer. On 12 July 1917 in addition to commanding Jagdstaffel 9, he was given command of the Jagdgruppe of the 3rd German Army.

By the end of the year, he was able to increase his number of aerial victories to five, and on 25 February 1918, he was transferred to Flieger-Ersatz-Abteilung 3 in Gotha as flight leader. On 14 June, he became head of the Experimental and Scientific Division at the command of the Flugmeisterei of Flieger-Abteilung A at Adlershorst, and six days later he was promoted to the rank of Hauptmann. He

Above: Oblt. Kurt Students Albatros D II with his personal marking, the emblem of the Armee-Ober-Kommando (A.O.K.) 3.

ended World War II as a general in the paratroops. He passed away in Lemgo on 1 July 1978.[6]

Vzfw. Erich Koehler, Albatros D II (L.V.G.), February 1917, Profiles 43, 43a, and 43b

Gen. ret. Student recalled that his personal markings were two crossed "Saracen swords" on a white square painted onto the fuselages sides. The color of the swords Gen. Student did not recall during the interview. Alex Imrie, who had interviewed Gen. Student ten years before me, was of the opinion that the swords were red. Therefore, I decided to go with the "red" variant. The L.V.G.-factory finish on this aircraft appears identical to the previous one.[7]

Information concerning **Vzfw. Erich Koehler´s** wartime career is very sparse indeed. All that is known about him is that he had joined the Saxon Army as a war volunteer. He was an pilot in the

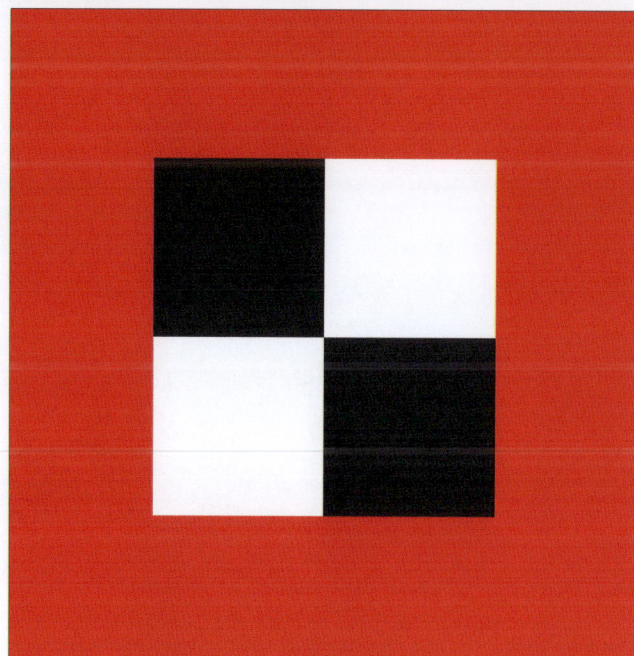

Above: The emblem of the Armee-Ober-Kommando served as the pattern for the personal marking that was applied to Oblt. Student's Albatros D II.

Above: The Albatros D II (LVG) with the A.O.K. badge takes off. The photo was taken by his mechanic Hans Mohr and vividly illustrated how well his personal marking was visible even from a distance.

Right: Staffelführer Oblt. Kurt Student.

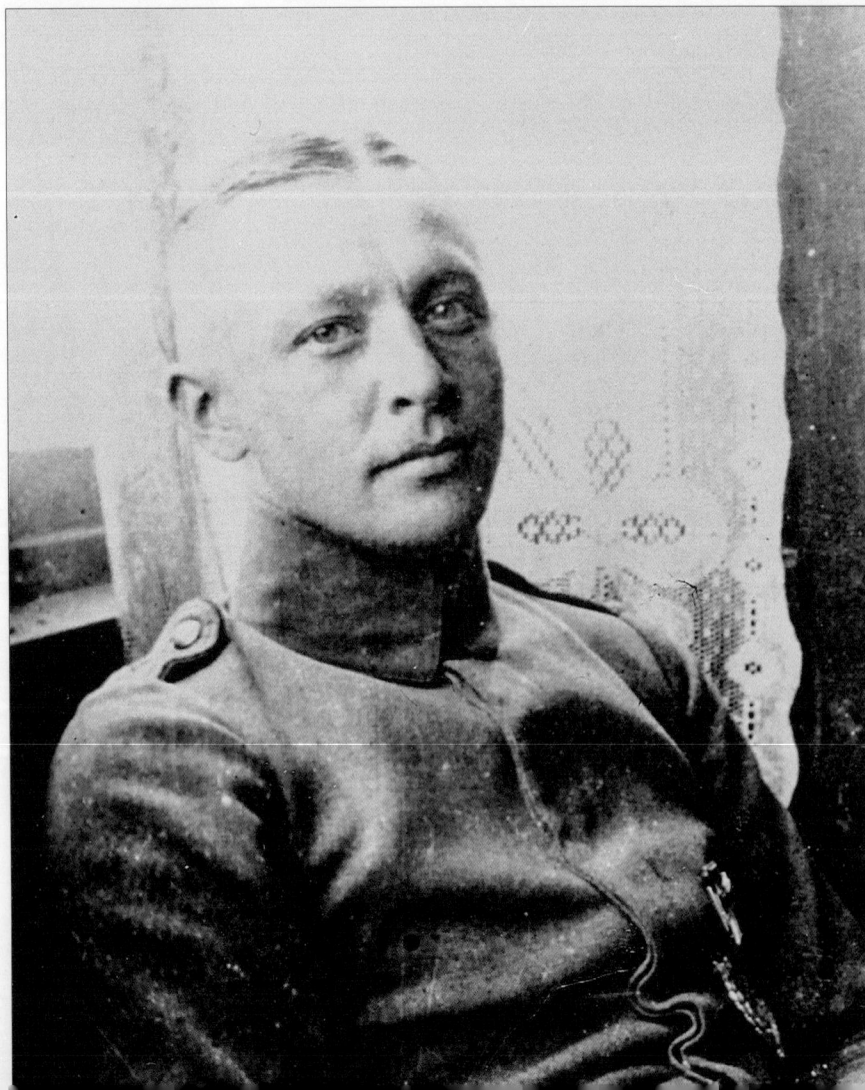

" Armeestaffel des A.O.K. 3" led by Oblt. Student and thus belonged to the nucleus of the newly established Jagdstaffel 9. On 27 May 1917, he was transferred from Jagdstaffel 9 to the Inspektion der Flieger. [8] Some records state that he crashed fatally at Chambry 30 May 1918. However, this was Lt. Karl Köhler. [9]

Vzfw. Erich Koehler,
Albatros D II (L.V.G.),
February 1917, Profiles
43, 43a, and 43b

Above: The Albatros D II of Vzfw. Erich Köhler. His personal marking was described as "Saracen swords". Note the mudguards that were fitted to the landing gear legs; a certain amount of rain apparently turned the landing ground at Leffincourt into a sludgy affair.

Below: Vzfw. Erich Köhler ready to take off with his Albatros D II (LVG) "Saracen swords". The upper wing camouflage patterns were almost "mirrored" on both sides on LVG-built examples of the Albatros D II.

Above: His Albatros D II parked next to a captured Caudron G 4. The large windscreen, rear-view mirror and flares box are interesting field additions to this aircraft that may be of interest to the modeler.

Right: Vzfw. Erich Köhler, of Jagdstaffel 9.

Above: Aircraft of Jagdstaffel 9 in early February 1917. In addition to the Albatros D IIs, the Staffel also received their first examples of the Albatros D IIIs around this time.

7.10 Royal Prussian Jagdstaffel 10

Staffelführer:

Oblt. Ludwig Link	10 October 1916 – 22 October 1916
Oblt. Helmuth Volkmann	6 November 1916 – 19 December 1916
Oblt. Rummelspacher	? December 1916 – 18 June 1917[1]

Until the 1970s, the period from Jagdstaffel formation in October 1916 to the winter of 1916/1917 was only sparsely documented in terms of written records. When Erich Tornuß visited the Reichsarchiv in Potsdam in the 1930s, the first part of the war diary of Jagdstaffel 10 covering the period from the formation of the Staffel to the end of February 1917 was missing. Accordingly, the war diary available in the Reichsarchiv did not begin until 1 March 1917.

The early history of Jagdstaffel 10 would have remained largely in the dark had it not been for Dr. Gustav Bock, who in the 1970s came across documents in the Bavarian Main State Archives containing information about the formation and operations of Jagdstaffel 10 in September/October 1916, which shed light on the Staffel's history for this period and which were used for this book.[2]

I received the most important personal information about the history and the painting schemes of the airplanes for the period autumn 1916–spring 1917 from the former fighter pilot Herrn Alois Heldmann and Herrn Xaver Leinemüller, a former mechanic of the Staffel.

Alois Heldmann was a member of Jagdstaffel 10 from its formation until the end of the war, something that was an extremely rare accomplishment. I visited him in Bad Aibling in the 1970s. He had just moved into a new apartment at that time and his photo album from his time with Jagdstaffel 10 was still packed in moving boxes. I agreed with him to visit him again at a later date to copy the photos and documents. Unfortunately, for some reason it did not come to that, and of course I still very much regret this today.

Around the same time, I also visited Herrn Xaver Leinemüller several times to interview him and to photograph his photo album. He had been transferred to the Staffel at the beginning of 1917 and belonged to the unit until the end of the war. During the conversations it became apparent how much knowledge the mechanics had about the airplanes and the painting schemes applied to them. After all, it was they who painted the airplanes of the Staffel according to the orders of the Staffelführer and the

Above: The second Staffelführer of Jagdstaffel 10, Oblt. Helmuth Volkmann, at far right as a member of Jagdstaffel 6 in October 1916. Next to him from left: Oblt. Max Reinhold and Rittm. Josef Wulff. (T. Weber).

Left: The third Staffelführer of Jagdstaffel 10, Oblt. Rummelspacher. He led the Staffel from the end of December 1916 to June 1917.

Facing Page: On the occasion of his visit in 1976, Bruno Schmäling was presented a drawing of Alois Heldmann's Fokker D VII, personally signed by him.

Above: Lt. Gustav Nernst in the cockpit of his Albatros D I which was marked with the red number "2". The formerly white background of the iron cross was covered with light green paint and represented the color of his parent unit, the Königlich Königlich preussisches Kürassier-Regiement "Graf Gessel (Rheinisches) Nr. 8 (Royal Prussian Cuirassier Regiment "Count Gessler" (Rhenish) No. 8). (M. Szigeti)

Above: Lt. Walter Bordfeld, in the cockpit of the Albatros D I "2" which he took over after Lt. Nernst was transferred to Jagdstaffel 30.

Right: The Albatros D I "2" at Chassogne-Ferme airfield in early 1917. The military number on the fin has been interpreted as 422/16 for the profile. When photographed, a replacement rudder appears to have been fitted.

Above: Albatros D II of Jagdstaffel 10 in March 1917 at Chassogne-Ferme airfield near Aincreville. The aircraft have been marked with red numbers on the fuselage sides for identification.

wishes of the pilots.

Very interesting photos for the period October 1916–March 1917 were found in the photo album of Gefr. Hermann Brettel, who was a fighter pilot with Jagdstaffel 10 from 26 May 1917, until he was wounded on 15 August 1917. In the 1970s I was able to locate his son, Herrn Reinhard Brettel. He made the photo album and the documents of his father available to me. When I photographed the photo album, I was very surprised to find a series of photos from the period Winter 1916/17 to March 1917, i.e. from a time when Hermann Brettel was not yet with the Staffel. He may have received these photos from comrades.

I received further interesting photos and documents from the relatives of Hans Junginger. He was one of Max Immelmann's mechanics and was transferred directly to Jagdstaffel 10 from Kampfeinsitzer-Kommando III on September 28, 1916. I also found two other very interesting photos from the formation period of Jagdstaffel 10 in the photo album of Carl Allmenröder.

When I worked on the book *"Royal Prussian Jagdstaffel 30"*, aviation historian Marton Szigeti informed me that he had some photos originating from Lt. Gustav Nernst. The photos not only cover his time with Jagdstaffel 30, but also includes a few photos of his time in Jagdstaffel 10, from the 25 September 1916 to January 1917.

During the period from 1 October 1916 to 20 October 1916, the following Albatros D I were assigned to Jagdstaffel 10, via the Armee-Flug-Park of the German 6th Army: Albatros D Is D.439/16, D.442/16, D.443/16, D.444/16, D.446/16, D.459/16, D.461/16, D.466/16 and D.468/16.[3]

On 28 October 1916, the day the Staffel was transferred to the German 5th Army, the Staffel received the first Albatros D IIs with military numbers D.472/16, D.488/16 and D.492/16 from Armee-Flug-Park 6.[4]

Photos of the Staffel at the airfield Chassogne-Ferme near Ancreville show the equipment of the Staffel with Albatros D I and Albatros D II aircraft. However, no exact date is available as to when the Staffel was moved there. The war diary transcript, dated 1 March 1917, only notes: "The Staffel is based at Chassogne-Ferme airfield near Ancreville." Neither Alois Heldmann nor Xaver Leinemüller

remembered when the move there occurred. They
vaguely remembered that the transfer must have
taken place sometime in the first weeks of 1917.

According to Herrn Heldmann and Herrn
Leinemüller, the Staffel received its first Albatros
D IIIs at the end of February 1917, as evidenced by
photographs showing Albatros D IIIs of the Staffel in
the snow.[5]

Furthermore, Herr Alois Heldmann reported
that Oblt. Volkmann introduced an identification
system consisting of numbers ranging from "1"
to "12" in order to identify the individual aircraft.
These numbers were painted in red on the plywood
fuselage.[6]

Lt. Gustav Nernst, Albatros D I, October 1916, Profile 44

The aircraft had a honey-yellow fuselage with a red
number "2" on both sides. Striking and unusual
is the Iron Cross located in a colored rectangle.
Likewise, the formerly white field of the Iron Cross
on the fin is also painted over. In Jagdstaffel 30, all
his aircraft had the color light green, sometimes
combined with white as a personal marking. This
is confirmed by the Report of the Flakgruppe-
Kommando in the Flugmeldebuch (aircraft spotting
book) of the German 6th Army from 3 April
1917. The color light green was a reference to his
parent unit, the Königlich preussisches Kürassier-
Regiement "Graf Gessel (Rheinisches) Nr. 8 (Royal

**Lt. Gustav Nernst, Albatros D I,
October 1916, Profile 44**

Prussian Cuirassier Regiment "Count Gessler" (Rhenish) No. 8), in which the uniform had medium green cap bands, collars, and cuffs.[7]

For this reason, it is more than likely that the colored field of the Iron Cross was light green.

Gustav Nernst was born on 25. January 1896 in Göttingen as the son of the later Nobel prize recipient in chemistry, Walther Nernst. He went to war with the Royal Prussian Cuirassier Regiment "Count Gessler" (Rhenish) No. 8. He transferred to the Fliegertruppe and after his training as a pilot he was transferred to Feld-Flieger-Abteilung 5 to fly single-seater fighter aircraft and was immediately assigned to Kampfeinsitzer-Kommando III. On 25 September 1916 he was taken over to the newly formed Jagdstaffel Link, and therefore become a member of Jagdstaffel 10.

On the 10 December 1916, he gained his first aerial victory when he shot down a French Nieuport single-seater at Esnes over French lines. On 20 January 1917 he was transferred to the newly formed Jagdstaffel 30. On 3 April 1917, after a dogfight, he forced Nieuport A6674 from 40th Squadron RFC to land inside German territory. The British pilot, 2/Lt. S A Sharpe, was captured unwounded. Two days later he was again successful. He forced Sopwith 1 ½ Strutter A1073 from 43rd Squadron RFC to land on the German side of the front. Together with the Staffelführer Oblt. Bethge he was regarded as the best fighter-pilot of the Staffel with a successful future. But his career soon come to an end. On 21 April 1917, Lt. Gustav Nernst, flying Albatros D III D.2147/16, was rammed by his comrade Lt. Oskar Seitz in a dogfight with F.E.´s and Sopwith Triplanes at 18:45 and crashed fatally over enemy territory

near Arras. Lt. Seitz was able to make a smooth emergency landing wiht his damaged aircraft in enemy territory near Gavrelle and managed to return on foot to his own lines. Lt. Oskar Seitz never got over this tragedy and was transferred to his former Fliegerabteilung (A)292 at his own request on 20 July 1917.[8]

Vzfw. Alois Heldmann, Albatros D II D.490/16, spring 1917, profiles 45

This Albatros D II D.490/16 was one of the first 50 Albatros D IIs ordered in August 1916 with the military numbers D.472/16–521/16. The fuselage of the aircraft had the usual "honey-colored" protective glaze. As a personal identification the aircraft had a red "12" on the fuselage.[9] A close analysis of the available photos by Jörn Leckscheid shows that the number had a narrow, considerably darker border. This confirms Mr. Heldmann's statement that the number was red and shows that it was a middle red. The edge would have been black. The upper wing surfaces had the factory finish rust red, light green and dark green camouflage paint, the lower wing surfaces were light blue.

The photos show the Albatros D II after an emergency landing, which Alois Heldmann rcalled during my visit:

"At that time (early 1917, author's note), the airfield had been severely softened by persistent rain and melt water. For this reason, a bulldozer had been used to prepare a takeoff and landing strip to such extent that it was possible to take off and land. However, if one deviated only a little from the leveled takeoff and landing runway, there was a great danger of getting stuck in the swampy meadow and ending up in a headstand with the

**Vzfw. Alois Heldmann, Albatros D II
D.490/16, spring 1917, profiles 45**

Above: Vzfw. Alois Heldmann crashed his Albatros D II D.490/16 in February 1917. As his individual marking he had the number "12" applied to the fuselage sides, which was given a very narrow outline in black.

Right: Magnification of the photo section in question shows the narrow dark outline around the number "12".

machine. During my landing my Albatros D II was caught by a sideways gust of wind, and I promptly put the crate down right next to the runway. Because of the crosswind I had opened the throttle again and landed with an appropriate speed in the wet meadow. Before I knew it, the Albatros D II had flipped over and landed on its back. I was unharmed except for a few bruises and crawled out from under the machine. My friend Adam (Barth), who was flying in the same group with me, and some

Above: Albatros D II of Jagdstaffel 10 in March 1917 at Chassogne-Ferme airfield near Aincreville. The aircraft have been marked with red numbers on the fuselage sides for identification.

Above: Albatros D II of Jagdstaffel 10 in March 1917 at Chassogne-Ferme airfield near Aincreville. The aircraft have been marked with red numbers on the fuselage sides for identification.

Above: Albatros D II of Jagdstaffel 10 in March 1917 at Chassogne-Ferme airfield near Aincreville. The aircraft have been marked with red numbers on the fuselage sides for identification.

mechanics came running. Someone had a camera with him and took pictures of my "ladies landing". I was heartbroken over my mishap, but that was not the last crash landing at the airfield.[10]

It remains to be explained in this context that landings where the aircraft lay down on its back were disrespectfully called "ladies' landings".

Alois Heldmann was born in Grevenbroich on 2 December 1895. At the outbreak of war at an engineering school, he volunteered for the front in January 1915 and initially fought as an infantryman in Russia. In the summer he transferred to the Fliegertruppe and joined Feldflieger-Abteilung 57 at the end of August. With this unit he flew in Serbia and Macedonia; later transferred to Feldflieger-Abteilung 59, in Flanders and on the Somme.

As a Vizefeldwebel, he retrained as a fighter pilot in the fall of 1916 and was transferred to Jagdstaffel 10 on 8 November. In May 1917, he achieved his first aerial combat success, but it was only recognized as "forced to land beyond the lines." In July 1917, Jagdstaffel 10 was incorporated into Jagdgeschwader I and in the same month he scored

his first two confirmed aerial victories.

Shortly thereafter he was promoted to Leutnant, and in November he scored his 3rd aerial victory. By the time the Staffel received its first Fokker D VIIs in late April 1918, his aerial victory count had increased to five. With the new aircraft, he more than doubled it over the next three months and was awarded the Knight's Cross of the Royal House Order of Hohenzollern around September 1918. By the end of the war, he was able to increase his number of aerial victories to 15. He was the only pilot of the Jagdgeschwader Freiherr von Richthofen No. 1 who had belonged to it from the first to the last day of its existence.

After the war he worked as a mechanical engineer and joined the then still secret German Air Force at the end of 1933. In 1938, as a Major in the Inspectorate of Flying Schools, he participated in the Deutschlandflug with a team of the Reichs-Luftfahrt-Ministerium. During World War II he was last as Oberst (Colonel) since 1 December 1942 Inspector of the Flying Schools. At the end of the war, he was taken prisoner by the Americans, but was released in 1946. At the end of the 1970s, he

Above: Alois Heldmann was a member of Jagdstaffel 10 from its formation until the end of the war.

Above: Vzfw. Adam Barth, close friend of Alois Heldmann in Jagdstaffel 10.

lived in Bad Aibling; he died in 1983.[11]

Vzfw. Adam Barth was one of the first to help his friend Vzfw. Alois Heldmann after the crash landing of the Albatros D II D.490/16. He also had himself photographed with Alois Heldmann in front of the crash-landed aircraft. However, a few days later, the same mishap happened to him. His plane was also caught by a sideways gust of wind. His plane still touched down at the edge of the runway, but also got caught in the swampy terrain and got stuck standing on its nose. This time it was up to Alois Heldmann to rush to his friend's aid, comfort him and take photos of the plane stuck in the swamp.

Adam Barth was born on 31 March 1896, in Weinheim in the Grand Duchy of Baden. On 25 March 1917, he reported the shooting down of a Nieuport south of Esnes, but it was not confirmed due to a lack of witnesses providing confirmation

of the crash from the ground. Shortly thereafter, on 3 April **1917**, he was promoted to Vizefeldwebel. On 5 December 1917, Vzfw. Barth achieved his first confirmed aerial victory. Under fire from his machine guns, a Bristol fighter crashed at 11:20 a.m. north of Cambrai. Less than two months later, on 30 January 1918, he crashed fatally at Anneux after aerial combat, at the controls of Albatros D V 4565/17.[12] According to Alois Heldmann, he had tried to come to the aid of another pilot who was being harassed by a British fighter plane and had himself been caught from behind by another British plane. In conversation with me Alois Heldmann remarked:

"I was deeply saddened by Adam's death. He was one of the best, most honest, and reliable friends I had had in the World War. The pain of such losses is not written in any war diary, one had to cope with it alone."[13]

7.11 Fighter Pilots in Two-Seater Units Autumn 1916/Winter 1917

A little-known story is the use of single or small groups of fighter planes assigned to two-seat units during the period autumn 1916 – winter 1916/ 1917 on the Western Front. The Jagdstaffeln established in the autumn of 1916 were initially primarily deployed to the main battle fronts, as with the German 1st, 2nd, and 5th Armies. Even though 24 German Jagdstaffeln had to been mobilized for service on the Western Front by the end of the year, this did not mean that there were already operational.[1] As a rule of thumb, it took one to two months after the mobilization of a Jagdstaffel before it was ready to be fully operational at the front. That Fliegerabteilungen, Fliegerabteilungen (A), or Kampfstaffeln could be protected by one or more fighter aircraft until enough Jagdstaffeln were operational at that front sector.

Based on available documents, the delivery of Halberstadt D II and D III aircraft to Feldflieger-Abteilungen 5 and 12, as well as to Fliegerabteilung (A) 280 is documented. However, it is likely that considerably more two-seater divisions were equipped with individual fighter planes until the corresponding number of Jagdstaffeln were ready for action at the front. In a number of cases, the basis for this was the Kampfeinsitzer-Kommando previously assigned to the Abteilung.

Two units are shown below as examples:

Kampfstaffel 11 of Kampfgeschwader II

For the depiction of the "Jagdstaffel" of Kampfstaffel 11, I was able to rely on the beautiful photos from the collection of Greg VanWyngarden, which he made available to me in his usual helpful manner. The written information can be found in the booklet "*Vom Jäger zum Flieger*" by Karl-Emil Schäfer.[2]

Lt. Karl-Emil Schäfer was transferred to Kampfstaffel 11 as a "combat single-seater pilot" on 18 November 1916.[3] In the winter of 1916/17, Kampfstaffel 11 received several Albatros D IIs, about which Karl-Emil Schäfer wrote home at the end of January 1917:

"Since recently, our Staffel has had a Jagdstaffel: von Hausen, Albert, Scheele, Herrmann, Stern and me."[4]

The "Jagdstaffel" of Kampfstaffel 11 existed for only about two months. After that, the fighter pilots of the "Jagdstaffel" were transferred to the "real" Jagdstaffeln which had recently been formed.

Only one of the six above-mentioned young fighter pilots, who enthusiastically climbed into the cockpits of their Albatros D IIs at Kampfstaffel 11 in the winter of 1916/1917 survived the year 1917, a fate they shared with a number of the early fighter pilots.

Lt. Karl-Emil Schäfer was born on 17 December 1891 as the oldest child and only son of a wealthy silk goods manufacturer in Krefeld-Bockum. He served his military service as a one-year volunteer in the years 1911–1912 with the Jäger Battalion No. 10 in Goslar and subsequently worked in London and Paris. In July 1914 he returned to Germany and at the outbreak of war was called up as a Unteroffizier to the Jäger Battalion No. 7, the renowned "Bückeburger Jägern". In September he received the

Above: Lt. Karl Emil Schäfer on the left as a member of Kampfstaffel 11 in the winter of 1916/17.

E.K. II and was promoted to Vizefeldwebel. Badly wounded in October 1914, he spent six months in hospital and then returned to the front with

Above: The five Albatros D II's of the "Jagdstaffel" of Kampfstaffel. From left: the Albatros D II's of: Lt. Herrmann, Lt. Albert, Lt. Schäfer, and Lt. von Hausen. An interesting detail can be noted on Schäfer's D.1724/16: rather than having his personal "disc" marking applied to the fuselage decking, it appears on the upper surfaces of the horizontal tailplane. (G. VanWyngarden).

his regiment. Promoted to Leutnant in May 1915, he enlisted in the Fliegertruppe six months later. Trained as a pilot from mid-January to the end of April 1916 at Flieger-Ersatz-Abteilung 8 in Graudenz and at the Köslin Flying School, he joined Kagohl II's Kampfstaffel 8 on 22 June. He spent the next few months flying in the front lines off Verdun and in Russia, retrained as a fighter pilot in late October, and was transferred to Kagohl II's Kampfstaffel 11 as a fighter pilot on 18 November 1916. The Staffel transferred to the Somme shortly before the end of

the year, where he achieved his 1st aerial victory on 22 January 1917.

On 21 February, via Armee-Flug-Park 6, he was transferred to Manfred von Richthofen's Jagdstaffel 11. He rose to the top of the German Jagdstaffeln within a few weeks. Seven aerial victories in March were followed by another 15 in April. During the course of that month, he first received the Knight's Cross of the Royal House Order of Hohenzollern and only a few days later, on 26 April 1917, he was awarded the Order Pour le Mérite after his

Above: Standing on the left wheel of Schäfer's D.1724/16, a mechanic conducts engine maintenance. The object placed on the ground, in front of the wheel, is the upper engine cover. Schäfer's personal marking on this aircraft was a white disc with a narrow black border. The marking represents a cannon ball, as he previously was a member of the artillery. The civilian visitor who has sneaked into the picture is unidentified.

Above: Lt. Werner Albert´s chevron-marked Albatros D II is seen here at a later date. Compared to the lineup view, the white square backgournd of upper wing crosses has been reduced, leaving just a white outline to the upper wing crosses. Oddly, this modification has not been carried out on the fuselage and tail. (G. VanWyngarden)

23rd aerial victory and at the same time appointed Staffelführer of Jagdstaffel 28. His series of successes continued with his new Staffel, and only a month later he had been credited with a totoal of 30 aerial victories. Just one day after his 30th kill, on 5 June 1917, flying alone ahead of his Staffel, he was

Above: In early February 1917, Jagdstaffel 31 was formed, and Lt. Werner Albert became the first Staffelführer of this unit. Here he is seated in the cockpit of an Albatros D III, likely during his time with Jagdstaffel 31. He was killed in aerial combat on 10 May 1917.

mortally wounded in aerial combat with F.E.'s and crashed, coming down near Zandvoorde at 4:05 p.m. and met his death.[5]

The available documents shows that he flew various Albatros D IIs with the following military numbers during his short time with Kampfstaffel 11: D.504/16, D.511/16 and D.1724/16. His personal insignia was a white disc outlined in black. After Schäfer's transfer to Jagdstaffel 11, Lt. von Scheele flew his Albatros D II D.504/16 for a short time.[6]

Lt. Werner Albert, Albatros D II, D.1707/16 or D.1776/16, winter 1916/1917, Profiles 46 and 46a (top view)

The aforementioned white chevron with the narrow black outline was applied to the fuselage sided and top decking of his personal Albatros D II. For the profile, a light-colored fuselage and the military number D.1707/16 have been chosen, both details are difficult do make out in the available photos. It is also possible that the plane had a dark-stained fuselage that appeared "rust-red" or "reddish-brown", and the military number may have been D.1776/16.

Lt. Werner Albert was born on 30 May 1885 in Düsseldorf. No information about his early military career is available. In January 1915 he transferred to the Fliegertruppe and was assigned to a Flieger-Ersatz-Abteilung, then to Kampfeinsitzerschule Freiburg (single-seat-fighter-school) to be trained as a pilot. In 1916 he was a pilot in Kampfgeschwader II. On 5 February 1917 was appointed Staffelführer of the newly formed Jagdstaffel 31. After scoring his sixth aerial victory on 10 May 1917, he crashed fatally during aerial combat near Vaudesincourt.[7]

His Albatros D II had as a personal insignia a white chevron with a narrow black outline on both sides and on the top of the fuselage. **When he was appointed Staffelführer of Jagdstaffel 31, he took this Albatros D II along to his new** Staffel. This was a fairly common practice that was probably allowed simply in order to save time. Pilots had a number of little personal modifications made to their planes, from adjustments to the seat and safety belt to the installation of flare holders and so on. They would not have to get these changes made to a new plane of identical type when they took along their "trusty old bird".[8]

Lt. Christian von Scheele was born on the 23 January 1895 in Schwerin. No documents are available about his early military career. After his service with Kampfstaffel 11, he was transferred to Jagdstaffel Boelcke on 31 January 1917. Soon after joining his new unit, during one of his first frontline flights with the new Staffel, he crashed fatally after aerial combat near Le Mesnil on 04 February 1917.[9]

His Albatros D II at Kampfstaffel 11 had a white "X" outlined in black on both sides and on the top of

Above: Lt. Christian von Scheele poses in front of his Albatros D II D.504/16. As can be seen from this very clear photo his black bordered white "X" marking was applied to the fuselage top, sides and bottom as well. (G. VanWyngarden)

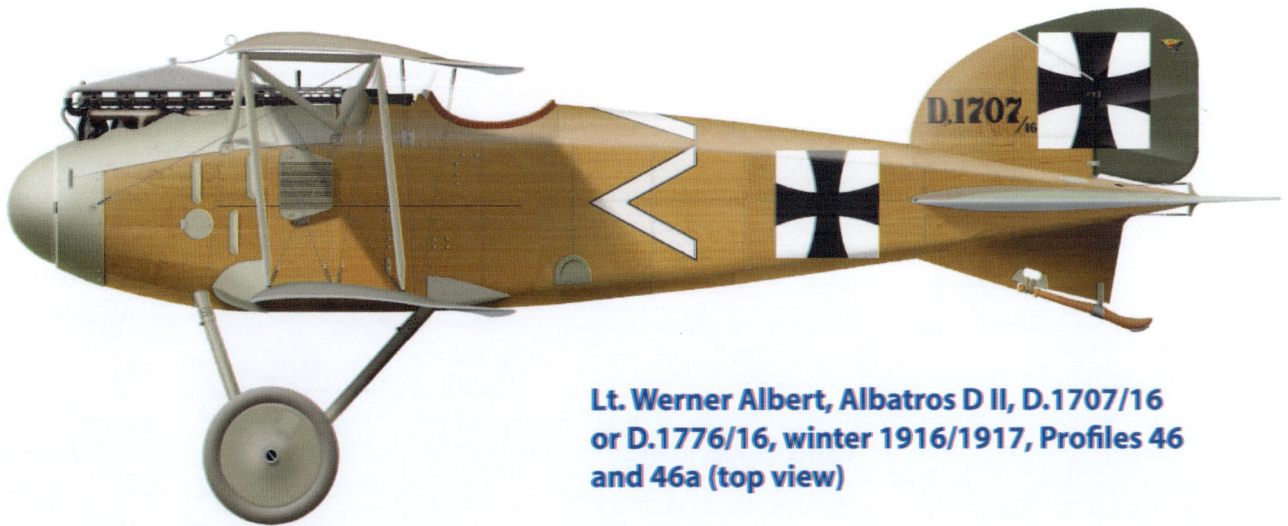

Lt. Werner Albert, Albatros D II, D.1707/16 or D.1776/16, winter 1916/1917, Profiles 46 and 46a (top view)

Above: Busy flight operations at the airfield of Kampfstaffel 11, with Lt. von Hausen's and Lt. Albert's Albatros D II in the foreground. The Iron Cross marking on the left side of the upper wing of Hausen's aircraft has been applied on a white square background, while the one on the right has a white outline. (G. VanWyngarden)

the fuselage as a personal insignia.[10]

Lt. Lothar von Hausen was born on 16 July 1894 in Leipzig. No documents are available about his early military career. After his service with Kampfstaffel 11 he was transferred to the new established Jagdstaffel 32 on 26 February 1917. On 16 March 1917, three Roland D IIs with the pilots Oblt. Schwandner, Lt. Rolfes and Lt. von Hausen took off on a frontline flight. Over French territory, a dogfight occurred with three SPAD VIIs of Escadrille SPA 3. In the process, Staffelführer Oblt. Schwandner was shot down in flames by Lt. Albert Deullin. Lt. Lothar von Hausen was seriously wounded and forced to land by Capt. Georges Gynemer. On 15 July 1917, he succumbed to his injuries in the military hospital.

Only Lt. Rolfes was able to save himself back across the German lines.[11]

His Albatros D II at Kampfstaffel 11 had a white triangular pennant with black trim on both sides and on top of the fuselage as a personal insignia.

Lt. Paul Herrmann was born in Hamburg on 10 November 1891. No documents are available about his early military career. On 20 February 1917, he followed his Staffel-mate Lt. Werner Albert, who had been appointed Staffelführer of Jagdstaffel 31, to this unit. On 19 April 1917, he crashed fatally after aerial combat with a SPAD fighter near Bois Nalval.[12]

According to Lt. Stern, his further career could not be ascertained.

Below: Lt. Paul Herrmann's Albatros D II was marked with a white wave line with a thin black outline on both sides, and the topside, of the fuselage. (G. VanWyngarden)

Fliegerabteilung (A) 280

The following account is based on the photo album and the personal information of Theodor Rumpel. He was one of the pilots with which I had much contact, and I visited him four times at his home at Büdingen. Details about his military career are in the book "*Royal Bavarian Jagdstaffel 23*", published by Aeronaut Books.

On 19 January 1917, he was transferred to Fliegerabteilung (A) 280, which was located at Schlettstadt Airfield in Alsace, in the area of Army Abteilung B. It was in early February when his career as a pilot took a sudden change, as he vividly recalled:

"*I was standing on the airfield on a cold but sunny day in early February 1917. I looked up and saw a single-seat Halberstadt D II floating towards our airfield. After the plane taxied out and the pilot disembarked, the sleek fighter was surrounded with great interest by the members of our Fliegerabteilung. This, then, was the fighter plane that would protect our heavy two-seat aircraft against enemy fighter attacks. The airplane seemed so small and light, almost fragile, but the elegant approach to our airfield alone had impressed me greatly. From this moment on I could not get the plane out of my mind.*

A few days later, I asked my Abteilungsführer, Hptm. von Holy if I could fly the single seater. After a short delay I received permission to fly the Halberstadt D II. Excitedly, I climbed into the pilot's seat of the Halberstadt. The pilot who had transferred the Halberstadt to our department had already explained the functions of the machine to me. I taxied carefully onto the runway. There I stopped again and checked all the functions of the plane most carefully, then I gave full throttle. The plane seemed to stay on the grass runway for only a few seconds, and it took off with an ease previously unknown to me.

I did a lap over the field and then soared up to fly south along the edge of the Vosges mountains. Nowhere was an enemy aircraft to be seen, the sky was clear except for some cumulus clouds. I pulled the plane higher and had fun dodging the cumulus clouds sideways, jumping over them or diving under them.

Flying an airplane had never seemed so easy, so light, so playful to me as with this airplane, which responded to every control movement, no matter how fine, which laid itself smoothly into turns, whereas the ponderous Rumpler [Rumpler C III; authors' note] could only be brought into a

Above: Fliegerabteilung (A) 280 Elsass, February 1916. Theodor Rumpel is far left, in the middle the Abteilungsführer Hptm. Holy.

Above: Theodor Rumpel's Halberstadt D II "Wolkenmaschine" (cloud aircraft) at Fliegerabteilung (A) 280. The patchy cloud finish was neatly applied to the fuselage, and a rectangular panel has been added to the fuselage below the exhaust in order to access the ammunition container. In addition to that, an insulating layer had been added to the central portion of the exhaust, most likely to prevent the pilot from suffering burns in case a gun jam needed to be cleared in the air.

Below: Front view of the Halberstadt D II "cloud aircraft" at Schleestadt airfield. A pointed metal cover has been mounted in front of the first cylinder of the engine, no doubt to ensure optimum operation of the powerplant during the cold winter months.

**Lt. Theodor Rumpel, Halberstadt
D II, January 1917, Profile 47**

sideways movement by pushing.

For me this flight was a revelation, it was like flying in a dream. Cocky, I tried tighter turns and steeper maneuvers, hoping to be far enough away from the airfield so that my Abteilungsführer would not become fearful for the Halberstadt. South of Mülhausen I turned the aircraft and flow via Colmar in a comfortable descent back to Schleestadt. The landing was soft and smooth. In my enthusiasm I had not noticed that it was actually still very cold.

From then on I tried to get permission to fly with the Halberstadt D II whenever possible. My desire to become a fighter pilot had awakened!"[13]

In the period that followed, he flew as many missions as possible with the Halberstadt D II, in addition to his usual duties at the controls of the Rumpler C III two-seaters, acting as a "guardian angel" for his comrades in the lumbering two-seaters. It was clear to him that as soon as the opportunity arose, he would apply for transfer to one of the Jagdstaffeln.

This said opportunity arose on a day in early March 1917. In the Alsace there was a period of "Fliegerwetter" (pilots´ weather), which meant that poor weather conditions did not permit flying activities. Along with several of his comrades, Theodor Rumpel drove to Colmar. There was a bar located in this city where the pilots and observers met. In the bar Theodor Rumpel got into conversation with another pilot, who turned out to be the Staffelführer of Jagdstaffel 26, Oblt. Bruno Loerzer. Both immediately got along with each other. Over a beer, Theodor Rumpel told his conversation partner about his test flights with the Halberstadt D II and his desire to become a fighter pilot. "Do you want to join me?" asked Bruno Loerzer and Theodor Rumpel answered the question with an enthusiastic:

"Yes!"

Bruno Loerzer got up and went to Hptm. von Holy, who was also sitting at a table in the bar, sat down with him and talked to him about Theodor Rumpel's wish to become a fighter pilot. After some time, Bruno Loerzer came back to Theodor Rumpel, patted Theodor on the shoulder and said to him:

"Everthing worked out fine, Hptm. von Holy has agreed to let you join my outfit. We will arrange for your transfer in the next few weeks.

Hptm. von Holy, who had also stood up and joined Theodor Rumpel at the counter, remarked with a sweet-sour face:

"That I can't hold you was clear to me from the first time you flew the single-seater. I wish you all the best and take good care of yourself."[14]

Lt. Theodor Rumpel, Halberstadt D II, January 1917, Profile 47

The factory finish of the aircraft was a sky-blue fabric and metal parts painted in the same color. Looking at the overall sky-blue aircraft, Theodor got the idea to paint clouds on it, as he told me:

"In contrast to the lumbering Rumpler (Rumpler C III), the Halberstadt felt as light as a cloud to me. Flying this crate was like floating on clouds. The plane was sky blue, so I had the idea to paint clouds on the fuselage of the plane. As a result, my Halberstadt quickly got the nickname "the cloud aircraft".[15]

Theodor Rumpel was born on 25 March 1897, in Bahrenfeld near Altona. On 1 October 1914, he had joined the Jäger-Regiment zu Pferde Nr. 6 in Erfurt as a war volunteer, deployed on the Eastern Front. At first, he still experienced exciting missions in reconnaissance rides and patrols, but the time of the cavalry soon came to an end on the Eastern Front as well. Promoted to officer in May 1916, he was transferred to Jäger-Regiment zu Pferde Nr. 2.

Since there was no longer any use for the cavalry, he enlisted in the Fliegertruppe and joined Flieger-Ersatz-Abteilung 3 in Gotha as a student pilot on 31 July 1916. From there he was transferred, via Armee-Flug-Park B, on 19 January 1917, as a pilot to Flieger-Abteilung (A) 280, where he was awarded the airman's badge as early as March 7. On 18 March 1917, he was assigned to Jagdstaffel 26, and less than a month later, on 12 April, he was transferred to Jagdstaffel 16, where he flew in a Ketto with Lt. Otto Kissenberth and Lt. Ludwig Hanstein.

After his friend Otto Kissenberth was appointed Staffelführer of Jagdstaffel 23, he made sure that Theodor Rumpel was transferred to his Staffel on 17 September 1917. By February 1918, he was able to increase the number of his aerial victories to five and on 26 February received the Bavarian Military Order of Merit IV. Class with swords. Four weeks later, on 24 March 1918, he was severely wounded in air combat over Bapaume, being wounded in the shoulder, and was sent to a military hospital. Transferred officially to FEA 2 at Schneidemühl on 13 April, he spent the following month in various hospitals where doctors were able to rescue his arm, but at the price of his shoulder remaining completely stiff.[16]

7.12 Summary

A summary of the available photos and documents of the first ten Jagdstaffeln shows that these single-seater units introduced personal identifications on their individual aircraft, for the most part, as early as late summer or early autumn 1916. In most cases, these markings appeared in the form of letters or numbers. But some early examples of fully painted fuselages, and the first "artistically applied" personal insignia could also be noted. Individual identifications and paint schemes also appeared on the fighter planes assigned for protection duties to the two-seater units.

Jagdstaffel 2, under the leadership of Hptm. Oswald Boelcke and Oblt. Stefan Kirmaier, stood out in a special manner, as a number of its aircraft had different colored fuselages as personal insignia as early as the autumn 1916. Even if Oswald Boelcke's aircraft did not have a personal insignia according to all we know to date, the pilots of his Jagdstaffel 2 can probably be regarded as the "fathers" of the colorful schemes of the German fighter planes.

What started off in the autumn of 1916 by the application of individual numbers or letters onto the fuselages of the aircraft evolved into the most colorful fighter planes that ever took part in a war in the following two years. Simple markings developed into downright picturesque works of art, and this can even be considered a small art form in its own right. But this is, to quote Rudyard Kipling: "another story", which I will present in the following volumes.

8. Factory Finishes of the First Halberstadt and Albatros Fighters

8.1 The Colors of the Factory Finish of Halberstadt Fighters

Analyzing the markings of the German Jagdstaffeln summer 1916 – winter 1916/1917, one must also keep in mind the factory schemes of these aircraft. This is especially interesting for modelers and artists.

To this day, no factory documents containing information about the color of the fabric covering the Halberstadt fighter planes has surfaced. These aircraft appeared in the orthochromatic black and white photos in the most diverse shades of gray from almost white (off-white) to dark gray. The color profiles of these **aircraft** shown in various publications are correspondingly diverse.

Alex Imrie and German aviation historian Heinz Nowarra had received information about the factory fabric covering from former pilots in the 1960s and 1970s. After Alex Imrie had presented the results of his research to me, I also began to ask my interlocutors, if they had flown this type of aircraft, to provide information on the factory finish of the covering of these aircraft.

Summarizing this information, the Halberstadt fighters had the following different fabric coverings, although this compilation does not claim to be complete:

8.1.1 Halberstadt Fighters with Brown Fabric

A number of Halberstadt D III and D V aircraft were covered with a brown-colored fabric. In his book *"El Shahin – der Jagdfalke"*, Oblt. Hans-Joachim Buddecke describes the front flight on 6 September 1916, and had the following to say about the Halberstadt D V:

"Again, we flew up and down over the front for half an hour. Here and there a dot appeared in the west, but it quickly disappeared at the sight of the first Kette on the German side.

Then [Oblt. Rudolf] Berthold took other direction, and we flew into enemy territory. Neat the captive balloons he dove down, we all followed suit, and soon I found myself having broken into a long line of "Großkodriger" [Grand

Above: One of the "brown rats" of Jagdstaffel 4, the Halberstadt D III of Vzfw. Ernst Clausnitzer at Roupy airfield. (J. Herris)

Caudrons] *practicing handling their fat charges.* [The Caudrons were practicing the protection of captive balloons, author's note]. *I immediately stood on my wing above the first one approaching me. He misunderstood this movement intended to make a turn in pursuit of me, but instead got right in front of my muzzle. I let the two machine guns hammer, and, smoking, he sped down into the deep. Next to me – the altimeter showed 900 meters – was a balloon decorated with cockades. I eased off, turned, and found myself facing a new "Großkodriger" from the same direction.*

The same was repeated a third time when I turned away from it... When my last one had sunk into the darkness of the clouds, I started on my way home. Now the Nieuports had to come soon; nothing could be done with the balloon without the ammunition intended for this anyway [Buddecke indicates that he did not carry phosphorous ammunition on this mission, author's note]. In the course of the battle, I had lost the man in front of me. As the last faithful, I saw Bernert with his brown biplane behind me."[1]

One more time, probably in October/November 1916, Hans-Joachim Buddecke mentions the color of the covering of Halberstadt fighter planes:

"*After half an hour or so, the enemy had to be completely alerted. The fighter field was swarming with cockades, all of which were fighters. It required our maximum effort to keep the upper hand over the enemy, especially since we were forced by our machines to always intercept them from below so as not to let them break through to the artillery and infantry planes . In the course of the movement, I met parts of mine again, who immediately attached themselves to me. Then we broke into the enemy anew. Again, and again the big (German) biplanes turned forward, who knew our **brown rats** well.*"[2]

These two statements made by Hans-Joachim Buddecke clearly indicate that the color fabric covering of the Halberstadt fighters flown by Jagdstaffel 4 was probably very similar to the brown-gray fur of a rat. (See profiles 35 and 36).

8.1.2 Halberstadt Fighters with Sky-Blue Covering

Rudolf Nebel recounted that his Halberstadt D III 127/16 at Jagdstaffel 5 was finished in an overall sky-blue scheme. He added that several of the Halberstadt D IIIs of Jagdstaffel 5 had the same light blue covering.[3]

Theodor Rumpel told me that the Halberstadt D II that was assigned to Fliegerabteilung (A) 280 and regularly flown by him had an oeverall light sky-blue scheme.[4]

The late German aviation historian Heinz Nowarra confirmed that numerous former pilots stated that some of the Halberstadt fighter planes serving with various Jagdstaffeln were covered with a light blue fabric. He published the result of this information in the book "*Eisernes Kreuz und*

Below: A Halberstadt D II covered with light blue fabric. The metal components and struts were apparently painted in a matching color, giving the aircraft an overall "sky blue" appearance. The man in the cockpit is Hermann Göring, photographed at Metz in June 1916.

Above: Lt. Josef Jacobs crash-landed this Halberstadt D V at the Valenciennes Jagdstaffel School. This aircraft is covered with a gray-blue fabric. It may have been one of the aircraft that were handed down to the Jagdstaffelschule from Jagdstaffel 5, although most Halberstadt fighters operated by this unit also had the numbers and letters applied to the fuselage decking as well.

Balkenkreuz"[5] (See Profile 38).

In this context, it is highly interesting to point out a document found by Reinhard Kastner in the Bavarian Main State Archives from the "Königlich Bayerischen Inspektion des Militär-, Luft- und Kraftfahrwesens" (Royal Bavarian Inspectorate of military aviation and motor transport). This states that: *"A light blue fabric is accepted for covering military aircraft"*.[6]

This document, when combined with the various available eyewitness accounts, confirms that the light blue fabric did indeed cover the fuselages and wings of a certain number of Halberstadt single-seat fighters. However, the currently available information cannot confirm if this fabric was used only on a single production batch of Halberstadt fighters or if was applied to the various sub-variants depending on availability of the fabric. All indications are that it was applied both to Halberstadt-built D II and D III fighters.

8.1.3 Halberstadt Fighters with a Darker Blue-Gray Covering

Among the photos of the Halberstadt D V fighters of Jagdstaffel 5 are several examples that appear comparatively dark in the orthochromatic photos. According to Rudolf Nebel, a number of the Halberstadt fighter planes used by Jagdstaffel 5 were covered with a darker blue-gray fabric.[7]

Aviation historian Heinz Nowarra confirmed to me that former fighter pilots who had flown Halberstadt single-seaters described this particular fabric covering as "dark blue-gray". According to these statements, a color profile of a Halberstadt fighter in this scheme was included in his book *"Eisernes Kreuz und Balkenkreuz"*.[8] (See Pofile 37)

8.1.4 Halberstadt Fighters with a Red-Brown and Green Camouflage, of Hannoversche Flugzeugwerke GmbH

The Halberstadt D II (Han.) aircraft built under license by Hannoversche Flugzeugwerke GmbH carried a reddish brown and medium green camouflage paint scheme. This was confirmed by the former Staffelführer of Jagdstaffel 25, Oblt. Friedrich-Karl Burckhardt, whose Jagdstaffel was equipped with Halberstadt D II (Han.), in conversation with Alex Imrie.[9] Alex Imrie included this account in his book *"German Fighter Units 1914–May 1917"*.[10]

Final Comment

Based on the available primary information, a sand-colored fabric covering of the Halberstadt aircraft, comparable to the one of the Fokker monoplane single-seaters, as it repeatedly appeared in publications in the past, has not been documented so far.

Above: Halberstadt D II, built under license by the Hannoversche Waggonfabrik AG, was covered in an overall rust-red - green camouflage paint scheme. The undersides were light blue.

Finally, the question remains why Halberstädter Flugzeugwerke G.m.b.H. used a variety of fabric covers for their single-seat fighters, which were built only in small numbers over a relatively short period of time. In 1916, large aircraft manufacturers such as Albatros, Aviatik or L.V.G. or were in a position to order large quantities of very specific fabrics from the fabric manufacturers due to the large numbers of aircraft they produced.

The relatively small Halberstadt Flugzeugwerke G.m.b.H. in 1916 was probably not in a position to order larger quantities of a particular fabric from one of the major manufacturers due to its production volume. It can be assumed that Halberstadt-Werke bought the fabrics currently available on the market and used them for its aircraft.

8.2 Factory Finishes of the Fuselages of the Albatros D I, D II, and D III

A careful analysis of the photos of the Albatros D I and D II show that two different coatings were applied to the plywood skin of the fuselages.

The apparently more common version was a clear varnish that seemed to have a "light honey" tinge to it. This resulted in an overall appearance that, for the sake of simplification, is referred to as "yellowish".

A darker version of the plywood fuselages has also been noted, and the color of these was reportedly "reddish brown" or "rust red". This color was most likely the result of a dark wood stain being used to

weatherproof the wooden surface.

When analyzing the photos, however, one must take into account a number of factors: These include the lighting conditions under which the photos were taken: the light conditions, the type of emulsion used for the development of the negative, the development of the photos, the type of paper of the photo itself, to name but a few of these factors.

Under certain lighting conditions, even fuselages that have been coated with the "yellowish" varnish can appear very dark and can be mistakenly

Above: Albatros D II (L.V.G. D I) 1044/16 of Jagdstaffel 23 with plywood fuselage coated with clear varnish that may have had a slightly yellowish tint, giving the aircraft an overall appearance that was described as "honey-colored" by German pilots in their post-war recollections. British and French reports described the color as "warm straw" or "light straw" in several cases. LVG did not apply the military number to the tail fin at the factory, in this case this was done at unit level.

interpreted as a rust-red stained example. (See *Jasta Colors* Volume 1, pages 255–256). As a basic rule, a reliable determination of the "color" of the wooden Albatros fuselage can only be made if aircraft with different coatings applied to their plywood skins have been photographed together and from the same angle.

8.2.1 Albatros D I and D II with "Yellowish" Varnish

According to the analysis of available photos, statements of former members of Jagdstaffeln to Alex Imrie and me, as well as the evaluation of official documents, the wooden fuselage of most of the Albatros-built D I, D II, and D III aircraft were coated with a shiny "yellowish" protective varnish, which was described by contemporary witnesses as "honey-colored", "warm straw" or "amber-colored" colored. In this context, the term "honey-colored" was referred to most often. L.V.G. built 75 Albatos D IIs under license, and photographs of these indicate that some had their plywood outer surfaces coated in

the varnish that gave the aircraft a "honey-colored" appearance. On others (such as the fighters operated by Jasta 9 illustrated in this book), apparently the darker wood stain was used. The L.V.G.-built Albatros D II was also termed L.V.G. D I in some official documents, but the former terminology has been used in this book in order to avoid confusion.

In the reports of ground troops and anti-aircraft commands, these aircraft are often referred to as having a "yellow fuselage". In most cases, these do not intend to state that the fuselage was painted yellow, but rather that varnished wood resulted in an overall yellow appearance when observed from the ground at a certain distance. However, it is recorded in some cases that the fuselages of Albatros fighters were actually covered in yellow paint![1]

8.2.2 Albatros D II and Albatros D III with "Reddish-Brown" Stain

In addition to the light-colored fuselages, a relatively small number of Albatros D II and D III aircraft

Above: Replica of Hptm. Oswald Boelcke´s Albatros D II 386/16 built by "The Vintage Aviator Limited" (TVAL). The aircraft made its first public appearance in 2013 and gives a very good impression of the "yellowish or honey-colored" appearance of the fuselage. (J. Herris)

are noted to have had dark fuselages. According to former German fighter pilots, these were aircraft which had their fuselages covered with a "reddish-brown" stain.[2] Former airmen also referred to the fuselages as "rust red." (See Profile 42 and 43). Unfortunately, no original fuselage component of such an Albatros D I/II/III is known to have survived. However, according to the airmen, the fuselage may have been more reddish than brownish. According to the state of knowledge based on available photos, the reddish-brown stain seems not to have been used on the Albatros D I. Current research indicates this dark weatherproofing of the plywood was used on some late-production D IIs and some early-production D IIIs.

It is not known why the Albatros Werke also used a reddish-brown stain on some aircraft they built in addition to the "honey-colored" varnish that was mainly used. A plausible explanation would be that, due to the rapid increase in orders, Albatros Werke were not able to secure a sufficient larger delivery of the varnish from their usual supplier. An entirely plausible explanation is that being forced to source an alternative "stain", that resulted in a darker appearance of the wooden fuselages.

In this context it should be noted that currently there is no evidence of the use of the "rust-red" varnish on the Albatros D V and D Va.

8.2.3 Albatros D II (O.A.W.) Fuselage Factory Finishes

In order to round off the short description of fuselage finishes, the fuselage-finishing practices of the Albatros Werke Schneidemühl ("A.W.S.", referred to in this book by the more commonly used term "O.A.W.") must be covered as well. It has to be pointed out that O.A.W. only produced a modest batch of 50 Albatros D II fighters. No factory records documenting the factory finishes of these aircraft is available to the authors, so only the small number of available photos of these aircraft serve as a source of interpretation.

The D IIs built by O.A.W. differed in some details to those manufactured by the parent company. These differences, as far as they could be determined, have all been faithfully reproduced in the relevant profiles in this book.

It seems rather surprising that available photos indicate that both the "honey-colored" varnish

Above: Albatros D II and Albatros D III of Jagdstaffel 4 in January/February 1917. The nose details of the aircraft nearest to the camera identify it as an O.A.W.-built Albatros D II (O.A.W.), and the dark fuselage indicates it may have a "rust red" wood-stained fuselage. Details of the D.II behind it do not permit certain identification of the manufacturer, but this is more likely an Albatros-built D II. Again, the fuselage appears dark enough to suggest a rust red" wood-stained fuselage. The Albatros D III behind these two aircraft and the D II have light "honey" colored fuselages.

can be noted (for example on aircraft delivered to Jagdstaffel 4), as well as the possible use of the "reddish brown" stain on some aircraft. This may be the case with the D.II (O.A.W.) carrying the number "2" on the side of its fuselage, as seen in Profiles 19 and 19a.

Finally, at least some O.A.W.-built D.IIs had their fuselages covered in a segmented camouflage scheme consisting of two or three colors: light green, dark green and/or reddish brown, in a similar manner to the wing upper surfaces. (See Profile No 39).

Above: An Albatros D III trio of Jagdstaffel 29 in the spring of 1917. The first machine, D.2185/16 marked "4", has a light "honey-colored" varnished fuselage. The two Albatros D III behind it, carrying the numbers "2" (D.2183/16) and "0" appear to be finished in a "rust-red" wood-stain.

Above: Albatros D II build by the Ostdeutsche Albatros Werke with the camouflage scheme.

Endnotes

Preface

1. Thor Goote, Rangehen ist alles, Berlin 1938
2. Extrapolation is a common method in research. It is an estimation of a value based on extending a known sequence of values or facts beyond the research area that is confirmed by primary data. In history, extrapolation is a method to project, extend, or expand known data or experience into an area not known or experienced so as to arrive at a usually conjectural knowledge of the unknown area.
The method of extrapolation is, of course, also speculative. But it is based on available first-hand information, and this information is applied it in a logical way to the unknown subject.
As an example: When we know that in a Jagdstaffel the fuselage bands on the aircraft of some pilots were references to the ribbons of the caps of their military arm or their parent unit, then it is a logical extrapolation that other aircraft with similar bands also presented the parent-units of the pilots. This extrapolation should be seen as valid until other first-hand evidence disproves it.

Prologue
Airfield of a German Jagdstaffel in Northern France in October 1916

1. "Vickers" or "Gitterschwänze" was a commonly used designation used by German aviators for all British aircraft in which the engine was located behind the cockpit and which were powered by a pusher propeller. A distinction was made between the "large Vickers" or "large lattice tails" such as the F.E.2b or the "small Vickers", or "small lattice tails" such as the Airco D.H. 2.
2. Kagohl, Kampfgeschwader der Obersten Heeresleitung (Fighting Squadrons of the Army High Command) for strategic bombing.

1. The German Fliegertruppe in 1914

1. Kriegswissenschaftliche Abteilung der Luftwaffe, Mobilmachung, Aufmarsch und erster Einsatz der deutschen Luftstreitkräfte im August 1914, Kriegswissenschaftliche Abteilung der Luftwaffe, S. 2, Berlin 1939
2. Neumann, Georg-Paul, Die deutschen Luftstreitkräfte im Weltkrieg, S. 64, Berlin 1920. Other sources give 228 aircraft, because Festungs-Fliegerabteilung 7 only had two aircraft.

3. Neumann, Georg – Paul, p. 64, ibid.
4. Hoeppner von, Ernst, Deutschlands Krieg in der Luft, p. 16, Leipzig 1921
5. Ernst von Hoeppner, Deutschlands Krieg in der Luft, p. 22, ibid.

1.1 Examples of the markings of the aircraft 1914

1. Mobilmachung, Aufmarsch und erster Einsatz der deutschen Luftstreitkräfte im August 1914, p. 119, ibid.
2. Mobilmachung, Aufmarsch und erster Einsatz der deutschen Luftstreitkräfte im August 1914, p. 111, ibid.
3. Color drawing of the DFW B.I B.451/14 captured by the French, archive Greg VanWyngarden.
4. Photo and notes in the photo album of Lt. Wilhelm Pier, Feldflieger-Abteilung 23, archive Marton Szigeti. As the victor's wreath is only partially visible, the victor's wreath on the profile is based on contemporary German postcards. Very likely, a comparable illustration also served as the inspiration for the marking on the aircraft, too.
5. Mobilmachung, Aufmarsch und erster Einsatz der deutschen Luftstreitkräfte im August 1914, S. 111, ibid.
6. Übersicht der Behörden und Truppen in der Kriegsformation, Teil 10 Luftstreitkräfte – Abschnitt B: Fliegerformationen, Verfügung des Chefs des Feldflugwesens W. Nr. 7775 Berlin 1919, Compiled by Reinhard Zankl in Propellerblatt, 8/2003.
7. Letter of Otto Parschau dated 25 May 1915 to Anthony Fokker, archive Bruno Schmäling.
8. www.frontflieger.de. Biographical data of members of the German air force. Thorsten Pietsch's website is highly recommended and offers a lot of information and photos about the German air force in WW1.
9. Winfried Bock, biographical compilation of German airmen of World War I with four or more aerial victories, unpublished manuscript. In the text, I deliberately refrained from distinguishing between active officers and reserve officers, as they played virtually no role in daily military life. This distinction is also sometimes omitted in official German documents. Following the destruction of many military documents during the Second World War, this distinction is no longer possible in a number of cases.
10. Übersicht der Behörden und Truppen in der Kriegsformation, Teil 10 Luftstreitkräfte

– Abschnitt B: Fliegerformationen, Kriegsministerium v. 13.09.14, Nr. 669/14. A7L, Bavarian Main Staate Archive, Dept. 4, Munich, compiled by Reinhard Zankl in Propellerblatt, 1/2001.

It is of course possible that Otto Parschau was briefly commanded to Feldflieger-Abteilung 42 during his affiliation to the BAO.

11. Winfried Bock, biographical compilation ibid.
12. Übersicht der Behörden und Truppen in der Kriegsformation, Teil 10 Luftstreitkräfte – Abschnitt B: Fliegerformationen, Kriegsministerium Nr. 1157/8.14. A7L Nr. 7775 Berlin 1919, ibid.
13. Contemporary color drawing of an Aviatik B in 1915 at Flieger-Ersatz-Abteilung 10 in Böblingen, archive, archive Greg VanWyngarden.
14. Contemporary postcards of Feldflieger-Abteilung 34, archive Reinhard Zankl.

2. The German Fliegertruppe 1915

1. Ernst von Hoeppner, Deutschlands Krieg in der Luft, p. 23
2. Georg Paul Neumann, p. 63, ibid.
3. Georg Paul Neumann, p. 63, ibid. Festungsflieger-Abteilungen (Fortress flying unit), Fliegerkorps der Obersten Heeresleitung (Flying Corps of the Army High Command).
4. Georg Paul Neumann, p. 64, ibid.

2.1 Examples of the Markings of the Aircraft 1914

1. Übersicht der Behörden und Truppen in der Kriegsformation, Teil 10 Luftstreitkräfte – Abschnitt B: Fliegerformationen, Bayerisches Kriegsministerium v. 1.11.14, Nr. 56011, ibid.
2. War diary of Feldflieger-Abteilung 7, compiled by Reinhard Kastner.
3. Kastner, Reinhard, Markierungen bei der Königlich Bayerischen Feldflieger-Abteilung 7 und der Königlich Bayerischen Fliegerabteilung (A) 293 in Das Propellerblatt No. 38, p. 26 – 35.
4. Gustav Diekmannshemke, Kriegsstammrolle 17961/35; 17962/19; 18153/52, Bavarian Main Staate Archive, Dept. 4, Munich, compiled by Reinhard Kastner
5. Übersicht der Behörden und Truppen in der Kriegsformation, Teil 10 Luftstreitkräfte – Abschnitt B: Fliegerformationen, Kriegsministerium Nr. 77xx 15 A7L, ibid.
6. *O'Connor, Neal, Aviation Awards of Imperial Germany in World War I*, Vol. IV, p. 174 – 176, additional personal information from the von Crailsheim family to Bruno Schmäling, ibid.
7. Reinhard Kastner, Flugzeuge der Armee-

Abteilung-Gaede 1915 in: Das Propellerblatt Nr. 2/2001, sowie Bund 35 Akte 1.6. Landwehr Division, Bavarian Main Staate Archive, Dept. 4, Munich.

8. Propellerblatt Nr. 2/2001, ibid.
9. Übersicht der Behörden und Truppen in der Kriegsformation, Teil 10 Luftstreitkräfte – Abschnitt B: Fliegerformationen, Bayerisches Kriegsministerium v. 26.9.14, 3668, ibid.
10. War diary of Feldflieger-Abteilung 9, compiled by Reinhard Kastner.
11. Georg Pfleiderer, Personalakte OP 9301, Kriegsstammrolle 17968/158, Bavarin Main Staate Archive, Dept. 4, Munich, complield by Reinhard Kastner.
12. Various letters from Fokker monoplane pilots to Anthony Fokker, archive Bruno Schmäling.
13. Übersicht der Behörden und Truppen in der Kriegsformation, Kriegsministerium vom 13.9.14, Nr. 669/9.14.A7L. ibid.
14. Übersicht der Behörden und Truppen in der Kriegsformation, Bayerisches Kriegsministerium vom 26.9.14, Nr. 668 I. ibid.
15. Knötel d.J., Herbert u.a., Das deutsche Heer, Friedensuniformen bei Ausbruch des Weltkrieges, Band III, Tafel 142

3. The German Fliegertruppe in 1916

1. Georg Paul Neumann, p. 64, ibid.
2. Übersicht der Behörden und Truppen in der Kriegsformation, Teil 10 Luftstreitkräfte – Abschnitt B: Fliegerformationen, Verfügung des Chefs des Feldflugwesens W. Nr. 22429, Berlin 1919, in Propellerblatt Nr. 8, Herbst/Winter 2003.
3. Georg Paul Neumann, p. 65, *ibid*.
4. Georg Paul Neumann, p. 65, *ibid*.
5. Weisungen für den Einsatz und die Verwendung von Fliegerverbände innerhalb einer Armee, Anlage 2, Großes Hauptquartier Mai 1917, Archive Bruno Schmäling
6. Weisungen für den Einsatz und die Verwendung von Fliegerverbände, ibid.
7. Personal Information by Josef Jacobs; FFA 11, Jasta 22, Staffelführer Jasta 7 to Bruno Schmäling.
8. Weisung für den Einsatz und die Verwendung von Fliegerverbänden innerhalb einer Armee, Anlage 2, *ibid*.

3.1 Examples of the markings of the aircraft 1916

1. Übersicht der Behörden und Truppen in der Kriegsformation, Kriegsministerium Nr. 82810.15A7L, *ibid*.
2. Photo album Ekkehard Reiss, AFA 211, Fokkerstaffel Falkenhausen, Archive Bruno

Schmäling.

3. Personal Information by Dr. Walter Böning, FFA 6b, Jagdstaffel 19, Staffelführer Jagdstaffel 76 to Bruno Schmäling.

4. Übersicht der Behörden und Truppen in der Kriegsformation, *ibid.*

5. War diary of Feldflieger-Abteilung 6b, compilation Reinhard Kastner.

6. A report in *"L'Aerophile"* describes the covering of a captured L.V.G. C II as follows: "The wings, tail and fuselage are all covered in white, or sometimes raw, light grey fabric, coated with transparent dope, occasionally tinted a very light blue. In conversation with former pilots, they described the covering of the L.V.G two-seaters in 1916 as either "white" or bluish", Archive Greg VanWyngarden.

7. Photo album and notes of Walter Böning, Bayerische Feldflieger-Abteilung 6, Archive Bruno Schmäling.

8. Mobilmachung, Aufmarsch und erster Einsatz der deutschen Luftstreitkräfte, *ibid.*

9. Photo album, documents, and notes of Carl Allmenröder, Feldflieger-Abteilung 18, Jagdstaffel 1, copy archive Bruno Schmäling.

10. Walter von Ebenhardt, Unsere Luftstreitkräfte, semi-official casualty list of the German air force, Berlin 1930.

11. Weekly activity report Stofl 6 from 25.09.1916, Bavarian Main Staate Archive, Dept. 4, Munich, compilation Dr. Gustav Bock, Archive Winfried Bock.

12. Winfried Bock, Biographical compilation, *ibid.*

13. Ranking list of the Armee-Flugparks 6, Bavarian Main Staate Archive, Dept. 4, Munich. Walter von Ebenhardt, Unsere Luftstreitkräfte, *ibid.*

14. Winfried Bock, biographical compilation, *ibid.*

15. Photos, documents, and notes of Franz Walz, Brieftauben-Abteilung-Ostende, Kagohl I, Staffelführer Jastas Boelcke, 19 und 34, Fliegerabteilung 304, Archive Reinhard Kastner. Waldemar Christensen, Kagohl 1, Jasta 5 und 46, Archive Bruno Schmäling.
Photos and documents of Rolf von Lersner, Kagohl I, Jagdstaffel Boelcke copy, Archive Bruno Schmäling.
Photos and documents of Hans von Keudell, Archive Greg van Wyngarden.

16. Reinhard Kastner, Markierungen – Kampfgeschwader / O.H.L. 1916, Das Propellerblatt Nr. 3, Frühjahr 2002.

17. Photo album and notes, Rolf von Lersner, Archive Bruno Schmäling

18. Letter from Herrn Dr. Albert von Lersner to Bruno Schmäling, 21 May 1982

19. Steffen Gastreich, Walter Waiss, Jagdstaffel Boelcke 1914 – 1918, p. 52, Aachen 2016.

20. Letter from Herrn Dr. Albert von Lersner, ibid. The semi-official casualty list of the German air force of Walter von Ebenhardt, in Unsere Luftstreitkräfte gives the 07.08.1917 as the date of death.

21. Alexis von Schoenermarck (Hrsg.): Helden-Gedenkmappe des deutschen Adels, Verlag Wilhelm Petri, Stuttgart 1921, p. 179

22. Personal information from Josef Jacobs to Bruno Schmäling.

23. Photo album, documents and notes from Walter von Bülow, Archive Bruno Schmäling.

24. Winfried Bock, Biographical compilation, ibid.

25. The single seater fighter, which were grouped together in loose commandos, appeared in the different armies under a wide variety of names. Since it was not an entitzed unit, there exists also no uniform designation.

26. Correspondence and personal information from Hans Sippel, mechanic of the Kampfeinsitzer-Kommando Ensisheim und Jagdstaffel 16 to Dr. Bock, 1968

27. Photos and documents of Fritz Grünzweig, Archive Tobias Weber.

28. Correspondence and personal information from Hans Sippel, *ibid.*

29. Friedrich Grünzweig Personalakte OP 20214, Kriegsstammrolle 18005/8
Compilation by Reinhard Kastner, War diary excerpt Jagdstaffel 16 by Erich Tornuß, revised and completed by Dr. Gustav Bock and Winfried Bock.

30. Personal Information by Gen. A.D. Otto Dessloch, bayerische FFA 9, KEK Ensisheim, Jasta 16, 35, Staffeführer Jasta 1 und 77.

4. The Formation of the First Ten German Jagdstaffeln 1916

1. War diary excerpt Jagdstaffel 3 by Erich Tornuß, revised and completed by Dr. Gustav Bock and Winfried Bock.

2. Handwritten notes by Alex Imrie based on his conversations with Johann Janzen, Kagohl 2, Kasta 12, Jasta 23, and Staffelführer Jasta 6.

3. Prof. Dr. Johannes Werner, Boelcke, p. 190, Leipzig 1942.

4. Chef des Generalstabes des Feldheeres, Weisung für den Einsatz und die Verwendung von Fligerverbände innerhalb einer Armee, Mai 1917, Archive Bruno Schmäling. There is still no clear definition of the term "air superiority". According to today's definition, air supremacy

also includes domination of enemy airspace, which the German air force on the Western Front, at least in the British sector, could hardly achieve. Therefore, the aforementioned mission is rather a declaration of intent – at least if one applies the modern definition. At best, it succeeded in achieving temporary air superiority over certain sections of ones own airspace, which might more accurately be termed air superiority.

5. Chef des Generalstabes des Feldheeres, Weisung für den Einsatz von Jagdstaffeln, 25.10.1917, Archive Bruno Schmäling.
6. Prof. Dr. Johannes Werner, Boelcke, p. 189–190, Leipzig 1942

4.1 Jagdstaffel 1
1. Übersicht der Behörden und Truppen in der Kriegsformation, Teil 10 Luftstreitkräfte – Abschnitt B: Fliegerformationen, Verfügung des Chefs des Feldflugwesens W. Nr. 22429, Berlin 1919.
2. War dairy excerpt Jagdstaffel 1 by Erich Tornuß revised and completed by Dr. Gustav Bock and Winfried Bock.
3. Photos of Jagdstaffel 1, Archive Alex Imrie and various other sources.
4. War diary excerpt Jagdstaffel 1. *Ibid.*
5. War diary excerpt Jagdstaffel 1. *Ibid.*

4.2 Jagdstaffel 2
1. Übersicht der Behörden und Truppen in der Kriegsformation, Verfügung des Chefs des Feldflugwesens W. Nr. 22429, ibid
2. War diary excerpt Jagdstaffel 2 transcription by Erich Tornuß, revised and completed by Dr. Gustav Bock and Winfried Bock.
3. War diary excerpt Jagdstaffel 2, ibid.
4. Prof. Dr. Johannes Werner, Briefe eines deutschen Kampffliegers an ein junges Mädchen, p. 65, Leipzig 1930

4.3 Jagdstaffel 3
1. Übersicht der Behörden und Truppen in der Kriegsformation, Verfügung des Chefs des Feldflugwesens W. Nr. 22429, ibid.
2. War diary excerpt Jagdstaffel 3, ibid.
3. War diary excerpt Jagdstaffel 3, ibid.
4. War diary excerpt Jagdstaffel 3, ibid.

4.4 Jagdstaffel 4
1. Übersicht der Behörden und Truppen in der Kriegsformation, Kriegsministerium vom 31.08.16 Nr. 929.16 g A7L, ibid.
2. War diary excerpt Jagdstaffel 4, transcription by Erich Tornuß.

3. War diary excerpt Jagdstaffel 4, transcription by Erich Tornuß.
   Lance Bronnenkant, *The Blue Max Airmen* Vol. 7, p.39, Aeronaut Book USA 2015
4. War diary excerpt Jagdstaffel 4, transcription by Erich Tornuß.
   Lance Bronnenkant, *The Blue Max Airmen* Vol. 3, p.109, Aeronaut Books, USA 2013.

4.5 Jagdstaffel 5
1. Übersicht der Behörden und Truppen in der Kriegsformation, Kriegsministerium vom 31.08.16 Nr. 929.16 g A7L, *ibid.*
   War diary excerpt Jagdstaffel 52 transcription by Erich Tornuß, revised and completed by Dr. Gustav Bock and Winfried Bock.
2. War diary excerpt Jagdstaffel 5, *ibid.*
3. War diary excerpt Jagdstaffel 5, *ibid.*
4. War diary excerpt Jagdstaffel 5, *ibid.*

4.6 Jagdstaffel 6
1. „Übersicht der Behörden und Truppen in der Kriegsformation, Verfügung Kriegsministerium vom 31.08.16 Nr. 929.16 g A7 & A.O.K. 5" dated 25 .08.1916. This source notes 23 August 1916 as the day of transforming Fokkerstaffel Sivry into Jagdstaffel 6, *ibid.*
2. War diary excerpt Jagdstaffel 6, via Erich Tornuß.
3. Nils Sörnsen, *Als Sänger-Flieger im Weltkrieg*, p.71, Hamburg 1933.
4. War diary excerpt Jagdstaffel 6, *ibid.*

4.7 Jagdstaffel 7
1. Übersicht der Behörden und Truppen in der Kriegsformation, Verfügung Kriegsministerium vom 31.08.16 Nr. 929.16 g A7. According to A.O.K. 5 records the day of transformation was 25.08.1916, *ibid.*
2. War diary excerpt Jagdstaffel 7, by Erich Tornuß, revised and completed by Dr. Gustav Bock and Winfried Bock.
3. War diary excerpt Jagdstaffel 7, *ibid.* Photo album of Wilhelm Eckenberg, Jagdstaffel 7, copy archive Bruno Schmäling.
4. War diary excerpt Jagdstaffel 7, *ibid.*

4.8 Jagdstaffel 8
1. Übersicht der Behörden und Truppen in der Kriegsformation, Verfügung Kriegsministerium vom 10.09.16, Nr. 24281 Fl., *ibid.*
2. War diary excerpt Jagdstaffel 8, by Erich Tornuß, revised and completed by Dr. Gustav Bock and Winfried Bock.

4.9 Jagdstaffel 9

1. According to information in the book "KEKs and Fokkerstaffels" there should have been a third Fokker-Staffel "Fokkerstaffel C" in Monthois. Unfortunately, no official documents relating to this matter have yet been found. Johan Ryheul, KEKs and Fokkerstaffels S. 143, Croydon 2014
2. War diary excerpt Jagdstaffel 9, by Erich Tornuß, revised and completed by Dr. Gustav Bock and Winfried Bock.
   Übersicht der Behörden und Truppen in der Kriegsformation, Verfügung Kriegsministerium Nr. 269.10.16 A7L, ibid.
3. Personal information from Gen. Kurt Student to Bruno Schmäling. Alex Imrie had interviewed Gen. Student in the 1960s and confirmed these statements.
4. War diary excerpt Jagdstaffel 9, *ibid*.

4.10 Jagdstaffel 10

1. War diary excerpt Jagdstaffel 10, by Erich Tornuß. The document „Übersicht der Behörden und Truppen in der Kriegsformation, Verfügung des Kriegsministerium Nr. 269.10.16 A7L", notes 28.09.1916 as the date of formation.
2. Weekly activity report Stofl 6 from from 10.10.1916, Bavarian Main State Archive, Dept. 4, Munich, compilation by Dr. Gustav Bock
3. Flugmeldebuch (flight report book) A.O.K. 6 vom 22.10.1916, Bavarian Main State Archive, Dept. 4, Munich

4.11 Summary

1. Alex Imrie, *German Fighter Units 1914–May 1917*, p.22, London 1978.
2. Übersicht der Behörden und Truppen in der Kriegsformation, *ibid*.

5. The Military Reasons for the Colorful Painting of the German Jagdstaffeln

5.1 Identification of the Aircraft of the Pilot's Own Staffel

1. Personal information from Josef Jacobs, ibid.
2. Summary of the statements of the mentioned Staffelführer to Bruno Schmäling, *ibid*.
3. Report and personal information of Walter Böning, Jagdstaffel 19 to the author based on the record of the altimeter, Archive Bruno Schmäling.
4. Personal information from Walter Böning, Jagdstaffel 19, *ibid*.
5. Personal information from Josef Jacobs, *ibid*. During his leadership from 27 December 1917 to 29 September 1918 Jagdstaffel 51 lost nine pilots

against the enemy and gained 15 aerial victories of which seven were gained by Oblt. Gandert.
6. Personal information from Josef Jacobs, *ibid*.
7. Personal information from Josef Jacobs, *ibid*.
8. Personal information from Josef Jacobs, *ibid*.
9. See Jasta Color Volume I, p. 258 – 263.
10. Personal information from Rudolf Nebel, Staffelführer KeSt. 1b.

5.2. Confirmation of Aerial Victories

1. This regulation did not yet apply to the period before the establishment of the Jagdstaffeln. Before that, it had happened in few individual cases that an aerial victory was awarded to several participants in the air combat.
2. Der Kommandierende General der Luftstreitkräfte, Nr. 580G 2, Weisungen für den Einsatz und die Verwendung von Fliegerverbänden innerhalb einer Armee, S. 12, Mai 1917, Archive Bruno Schmäling.
3. War diary excerpt Jagdstaffel 31, by Erich Tornuß, revised and completed by Dr. Gustav Bock and Winfried Bock, Archive Bruno Schmäling.
4. War diary excerpt Jagdstaffel 73, by Erich Tornuß, revised and completed by Dr. Gustav Bock and Winfried Bock, Archive Bruno Schmäling.
5. Personal information from Josef Mai to Bruno Schmäling.
6. Personal information Josef Jacobs, *ibid*.
7. Information by French aviation historian David Méchin.

6. The Letters and Number Markings of the First German Jagdstaffeln

1. Even though color reproduction in photographs was experimented with before World War I, almost all known photos of World War I are in black and white. An exception available to the authors are a few color photos of German booty planes that were exhibited in Paris.
2. Summary of the information from: Walter Böning Jagdstaffel 19, Hermann Dahlmann, Jagdstaffel 29, Alfred Fleischer Jagdstaffel 17, Otto Fuchs Jagdstaffel 30, Heinz Geisseler, Jagdstaffel 33, Alois Heldmann Jagdstaffel, Josef Jacobs, Jagdstaffel 22, Ludwig Marchner, Jagdstaffel 32, Josef Mai Jagdstaffel 5, Rudolf Nebel Jagdstaffel 5, Matthäus Wiest, Jagdstaffel 32, Ferdinand Zilcher, Jagdstaffel 1.
   Information from Alex Imrie about his conversation with Hans-Hermann von Budde, Jagdstaffel 29 und 15, Gerhard Fieseler, Jagdstaffel 25, Karl-Friedrich Burckhardt, Jagdstaffel 25, Karl Treiber, Jagdstaffel 5.
3. Siehe *Jasta Colors* Volume I, p.245–257

7. The Marking and Painting of the First 10 German Jagdstaffeln Autumn 1916 – Winter 1916/1917

1. Bruno Schmäling, Winfried Bock, *Royal Prussian Jagdstaffel 30, ibid*. Persönliche Auskunft Otto Fuchs

7.1 Königlich Preußische Jagdstaffel 1
1. War diary excerpt Jagdstaffel 1, ibid.
2. Winfried Bock, Biographical compilation, ibid.
3. Ed Ferko „Fliegertruppe 1914 – 1918". Photo-album Franz Ray and Raimund Armbrecht, Jagdstaffel 1, copy archive Bruno Schmäling.
4. Photo-album Raimund Armbrecht, photos Franz Ray.
5. Personal information from Rudolf Nebel, *ibid*.
6. Winfried Bock, Biographical compilation, *ibid*.
7. https://de.wikipedia.org/wiki/Hanns_Braun_(Leichtathlet).
8. Personalakte OP 5780 Hanns Braun KrStR 18012/41 Hanns Braun
9. Flight log Oblt. Hubert Greim, copy Archive Bruno Schmäling.
10. KrStR 18012/41 Hanns Braun.
11. https://www.hall-of-fame-sport.de/mitglieder/detail/Hanns-Braun.
12. The Albatros D I D.434/16, D.436/16 and 438/16 from this shipment were delivered to Jagdstaffel 2, Akte KA Iluft 204, Bavarian Main State Archive, Dept. 4, Munich

In Trichtern und Wolken, Adolf Ritter von Tutschek, Kriegsaufzeichnungen, p. 158, Braunschweig 1934, as well as *Over the Front* Volume 3, Number 4.
13. Personal statement by Franz Ray, to Alex Imrie. The claim that this aircraft was flown by a Lt. Karl Spitzhoff of Jagdstaffel 5, which can be read over and over again in various publications, does ot accord with the facts. Lt. Spitzhoff can't be found in the unit records of Jagdstaffel 5. Also, Jagdstaffel 5 was uniformly equipped with Halberstadt D III/D V at that time and never received Albatros D Is. The personnel roster of Jagdstaffel 1 does not carry a pilot by the name Spitzhoff. From 8 October 1916 to 15 December 1916 Lt. Karl Spitzhoff was a member of Jagdstaffel 6. However, according to available documents the Albatros D Is operated by this squadron had numbers as identification markings and no symbols.
14. Personal information from Otto Fuchs, *ibid*. Bruno Schmäling & Winfried Bock, *Royal Prussian Jagdstaffel 30*.
15. Personal information from Otto Fuchs, *ibid*.

16. Bruno Schmäling & Winfried Bock, *Royal Prussian Jagdstaffel 30*.
    A.O.K. 6 Flieger Nachrichten- und Verfolgungsstelle (6. Army Department of information and research) as well as Rangliste Armee-Flugparks 6, Bavarian Main Staate Archive, Dept. 4, Munich.
17. In 1911, Ludwig Sütterlin was commissioned by the Prussian Ministry of Culture and Education to develop an initial script for learning cursive writing in school. From 1915, the German Sütterlin script was introduced in Prussia. Many handwritten reports of the 1st World War are written in this script and are correspondingly difficult to read.
18. War diary excerpt, Jagdstaffel 1, ibid. Additional information by von Greg VanWyngarden about the 28 December 1917 aerial combat.
19. Personalakte 6/3092 Raimund Armbrecht, Bundesarchiv Freiburg.

7.2 Königlich Preußische Jagdstaffel 2
1. War diary excerpt Jagdstaffel 2, *ibid*.
2. See also Jasta Color Volume I, S. 33–34.
3. Handwritten notes of Herbert Schulz, according to statements of former members of the Jasta Boelcke in the 1920s.
   See also *Jasta Colors* Vol. 1, p.33 – 34, ISBN 978-1-953201-00-3, Aeronaut Books, 2020.
4. The Albatros D I and D II carried the military number on the tail fin. The Albatros D IIs built under license by L.V.G. and O.A.W. were an exception, as no military number was applied to the tail fin at the factory. According to available documentation, Jagdstaffel 2 was not equipped with any L.V.G- or O.A.W.-bulit Albatros D IIs. Accordingly, all Albatros D IIs should have carried the serial number on the tail fin. If this is not visible the only logical conclusion is that it was overpainted.
5. See also: Photos in Lance Bronnenkant, *The Blue Max Airmen* Vol. 1 and Vol. 5, as well as *Oswald Boelcke, The Red Barons Hero*, Aeronaut Books, USA 2018. Both books are highly recommended!
6. Winfried Bock, Biographical compilation, *ibid*. Lance Bronnenkant, *The Red Barons Hero, ibid*.
7. War diary excerpt Jagdstaffel 2, *ibid*.
8. Personal information from Herr Franz Piechulek, Jagdstaffel 20, KeSt. 5, Jagdstaffel 41 und 56 to Bruno Schmäling,
9. Bruno Schmäling, *Jasta Colors* Volume 1, Grey Scale Interpretation, p.255–256, *ibid*.
10. Combat in the Air, No 60 Squadron, 27.12.1916. Copy G. VanWyngarden Collection.
11. Hptm. Adolf Ritter von Tutschek, *Stürme und*

Luftsiege, p. 95, Berlin 1918

12. Winfried Bock, Biographical compilation, ibid.

13. Hptm. Adolf Ritter von Tutschek, *Stürme und Luftsiege*, p. 95, Berlin 1918.

14. Herbert Knötel d.J., Paul Pietsch, Baron Collas Friedrich Herrmann, Georg Ortenburg, Ingo Prömper, Hans Rudolf von Stein, Das Deutsche Heer, Friedensuniformen bei Ausbruch des Weltkrieges, Board 85–86, Stuttgart, 1982.

15. Hptm. Adolf Ritter von Tutschek, *Stürme und Luftsiege*, p. 94–98, Berlin 1918. The report of Adolf von Tutschek in the later published book *"In Trichtern und Wolken"* differs from this report in details.

16. Short biography about Friedrich Karl von Preußen, compilation by Lance Bronnenkant. Die Helden-Gedenkmappe, lists 6. April 1917 as the date of death.

17. Herbert Knötel u.a. *Das deutsche Heer*, Band II, S. board 48, ibid.

18. An example: Lt. Waldemar von Buttlar of FFA 22 and Theodor Rumpel, of Jagdstaffel.

19. Winfried Bock, Biographical compilation, *ibid*. Lance Bronnenkant, *Blue Max Airmen* Vol. 11, S. 108, Aeronaut Books 2018, Prof. Dr. Johannes Werner, *Briefe eines deutschen Kampffliegers an ein junges Mädchen*, Leipzig 1930.

20. Manfred von Richthofen, *Der rote Kampfflieger*, S. 108, Berlin- Wien 1917. Manfred von Richthofen had long rejected offers from publishers to write a book. Under pressure from the Supreme Army Command, he finally agreed and signed a contract with the Ullstein Verlag. The contract with Ullstein stipulated that the journalist Erich von Salzmann should revise the manuscript not only linguistically and stylistically, but also in terms of content. This journalist ensured that the expectations of the Supreme Army Command and the readers were met. Considering factual and chronological inconsistencies in the book, von Salzmann must have changed the original text by quite a bit. From today's point of view, it is therefore impossible to determine what von Richthofen wrote by himself and what was changed, rewritten, or simply added by von Salzmann. For this reason, the book is to be regarded with skepticism. To analyze Richthofen's personality based on the statements in this book is therefore doomed to failure from the beginning. An example for such a failure is the book *"Der rote Kampfflieger von Rittmeister Manfred Freiherrn von Richthofen"* by Friedrich-Wilhelm Korff, published in Germany 1977, who tried to portray the personality of Manfred von Richthofen based

on this book.

21. Herbert Knötel d.J.u.a., *Das deutsche Heer*, board 100, *ibid*.

22. Combat in the Air, No 60 Squadron, 27.12.1916.

23. Akte MMJO V K 14 / 10, Max Ritter von Müller, Bavarian Main Staate Archive Dept. 4, Munich, compilation by Reinhard Kastner.

24. **Handwritten war diary excerpt of Jagdstaffel Boelcke, by Erich Tornuß. Archive Bruno Schmäling.**

7.3 Königlich Preußische Jagdstaffel 3

1. War diary excerpt Jagdstaffel 3, *ibid*.

2. War diary excerpt Jagdstaffel 3, *ibid*.

3. War diary excerpt Jagdstaffel 3, *ibid*.

7.4 Königlich Preußische Jagdstaffel 4

1. War diary excerpt Jagdstaffel 4, *ibid*.

2. War diary excerpt Jagdstaffel 4, *ibid*. Fotos in the Album of Ernst von Althaus.

3. Hans-Joachim Buddecke, *El Schahin – Der Jagdfalke*, p.94–95 and p.103, Berlin 1918.

4. Winfried Bock, Biographical compilation, *ibid*.

5. Winfried Bock, Biographical compilation, *ibid*.

6. Personal information from Herrn Xaver Leinemüller, mechanic of Jagdstaffel 10 to Bruno Schmäling, confirmed by Herrn Alois Heldmann, Jagdstaffel 10.

7. Compilation based on the research of Herrn Dr. Hannes Täger, see also Hannes Täger, The Lost Honor of Ernst Freiherr von Althaus, in *Over the Front*, Volume 29, Number 2, Summer 2014, p.100–147.

8. Handwritten notes of Alex Imrie according to his interviews with Alfred Lenz.

9. Hannes Täger, *ibid*.

10. Hannes Tager, *ibid*.

11. This is not a definitive statement. Photos of the Albatros D II may well turn up with different personal markings.

7.5 Königlich Preußische Jagdstaffel 5

1. War diary excerpt Jagdstaffel 5, *ibid*.

2. For detailed information about Rudolf Nebel see *Jasta Colors* Vol. 1, p.105–114.

3. Personal information from Rudolf Nebel, *ibid*.

4. Personal information from Rudolf Nebel, *ibid*.

5. Winfried Bock, Biographical compilation, ibid.

6. Personal information from Rudolf Nebel, *ibid*.

7. Winfried Bock, Biographical compilation, ibid. Michael Stacey, Hans Müller ein erstklassiger Eindeckerflieger, in *Cross & Cockade* Vol. 24, No. 2, Summer 1983, p.97–108.

7.6 Königlich Preußische Jagdstaffel 6
1. War diary excerpt Jagdstaffel 6, *ibid.*
2. Niels Sörnsen, *Als Sänger-Flieger im Weltkrieg*, Carl Holler-Veralg,1933.
3. War diary excerpt Jagdstaffel 6, *ibid.*
4. Peter Grosz, *Albatros Fighters*, p.2 und 55., Berkhamsted, 1991.
5. Analysis of photos of the different Albatros D II variants by Jörn Leckscheid.
6. War diary excerpt Jagdstaffel 6, Nachrichtenblatt der Luftstreitkräfte.
7. Niels Sörnsen, *Als Sängerflieger im Weltkrieg*, *ibid.*
8. Niels Sörnsen, *Als Sänger-Flieger im Weltkrieg*, p.75–76, *ibid.*

7.7 Königlich Preußische Jagdstaffel 7
1. War diary excerpt Jagdstaffel 7, *ibid.*
2. War diary excerpt Jagdstaffel 7, Photo-albums of Lt. Wilhelm Eckenberg, Vzfw. Friedrich Mannschott and Oberflugmeister Kurt Schönfelder.
3. Rangliste Jagdstaffel 32, Bavarin Main Staate Archive, Dpartment 4, Munich.

7.8 Königlich Preußische Jagdstaffel 8
1. War diary excerpt Jagdstaffel 8, *ibid.*
2. Documents and research by Greg VanWyngarden and Lance Bronnenkant, made available to this book.
3. Winfried Bock, Biographical Compilation, *ibid.*

7.9 Königlich Preußische Jagdstaffel 9
1. War diary excerpt Jagdstaffel 9, *ibid.*
2. Original document of aircraft and aircraft engines used by the flying units of the German 3rd Army in October 1916, archive Bruno Schmäling.
3. Photos Hans Mohr, mechanic of Jagdstaffel 9, confirmed by Gen. Kurt Student to Bruno Schmäling.
4. Personal information from Gen. Kurt Student to Bruno Schmäling.
5. Personal information from Gen. Kurt Student, *ibid.*
6. Winfried Bock, Biographical compilation, *ibid.*
7. Personal information from Gen. Kurt Student to Alex Imrie.
8. Research of Hannes Täger about Erich Koehler.
9. War diary excerpts Jagdstaffel 9, *ibid.*

7.10 Königlich Preußische Jagdstaffel 10
1. War diary excerpts Jagdstaffel 10 by Erich Tornuß, revised and completed by Dr. Gustav Bock and Winfried Bock.
2. Armee-Tagesbefehl 21.09.1916, Stabsoffizier der Flieger 6, wöchentlicher Tätigkeitsbericht Stofl 6 vom 25.09.1916, Bavarian Main Staate Archive Dept. 4, Munich
3. KA ILUFT204 - Flugzeuge Armeeflugpark 6, La Briquette, 31.10.1916, Bavarin Main Staate Archive, Dpartment 4, Munich.
4. KA ILUFT204 - Flugzeuge Armeeflugpark 6, La Briquette, 31.10.1916, *ibid.* Photo-album Hans Junginger, Hermann Brettel, both Jagdstaffel 10, copy Archiv Bruno Schmäling.
5. Personal information from Alois Heldmann and Xaver Leinemüller. *ibid.*
6. Personal information from Alois Heldmann, *ibid.*
7. **Flugmeldebuch (report on air activity) of the German 6th Army, report of the Flakgruppen-Kommando, Bavarian Main State Archive, Department 4, Munich.**
8. Bruno Schmäling, Winfried Bock, *Royal Prussian Jagdstaffel 30*, Aeronaut Books.
9. Personal information from Alois Heldmann, *ibid.*
10. Personal information from Alois Heldmann, *ibid.*
11. Winfried Bock, Biographical Compilation, *ibid.* Personal information from von Alois Heldmann *ibid.*
12. War diary excerpt Jagdstaffel 10, *ibid.*
13. Personal information from Alois Heldmann, *ibid.*

7.11 Jagdflieger bei Doppelsitzer-Einheiten
1. Übersicht der Behörden und Truppen in der Kriegsformation, Teil 10 Luftstreitkräfte – Abschnitt B: Fliegerformationen, *Ibid.*
2. Lt. Schäfer, *Vom Jäger zum Flieger*, Tagebuchblätter und Briefe, p. 77, Berlin 1918.
3. Rangliste des Armee-Flug-Parks 6, Bavarian Main Staate Archive Dept. 4, Munich
4. Lt. Schäfer, *Vom Jäger zum Flieger, Tagebuchblätter und Briefe*, p. 77, *ibid.*
5. Winfried Bock, Biographical compilation, *ibid.*
6. Lt. Schäfer, *Vom Jäger zum Flieger*, p.78 ff., *ibid*, Lance Bronnenkant, *Blue Max Airmen*, Vol. 7, p.59 ff. Photos and documents from Greg VanWyngarden.
7. War diary excerpt Jagdstaffel 31, *ibid.*
8. War diary excerpt Jagdstaffel 31 *ibid.* Personal information from Fritz Jacobsen, Jagdstaffel 31 to Bruno Schmäling.
9. War diary excerpt Jagdstaffel Boelcke, *ibid.*
10. Lt. Schäfer, Vom Jäger zum Flieger, p. 79, *ibid.* Lance Bronnenkant, *Blue Max Airmen*, Vol. 7, p.60. Photos and documents from Greg VanWyngarden.
11. War diary excerpt Jagdstaffel 32 , *ibid.* David Méchin, *The WW I French Aces Encyclopedia*, Volume 4, p.178, Aeronaut Books 2021.
12. War diary excerpt Jagdstaffel 31, *ibid.* Walter von

Ebenhardt, *Unsere Luftstreitkräfte*, semi-official casualty list of the German army air services 1914–1918.

13. Personal information from Theodor Rumpel to Bruno Schmäling, 1977.

14. Bruno Schmäling, Winfried Bock, *Royal Bavarian Jagdstaffel 23*, p.8, *ibid*.

15. Personal information from Theodor Rumpel to Bruno Schmäling, 1977.

16. Winfried Bock, Biographical compilation, *ibid*. Personal information from Theodor Rumpel, *ibid*. Bruno Schmäling, Winfried Bock, *Royal Bavarian Jagdstaffel 23*, Aeronaut Books.

3. Personal information from Rudolf Nebel, *ibid*.
4. Personal information from Theodor Rumpel, *ibid*.
5. Heinz J. Nowarra, *Eisernes Kreuz und Balkenkreuz*, p.130, Berlin 1968.
6. Akte Iluft 80, Königlich Bayerische Inspektion des Militär-,Luft- u. Kraftfahrwesens, Bavarian Main Staate Archive, Dept. 4, Munich.
7. Personal information from Rudolf Nebel, *ibid*.
8. Heinz J. Nowarra, *Eisernes Kreuz und Balkenkreuz*, p.130, *ibid*.
9. Handwritten notes of Alex Imrie about the interview with Friedrich-Karl Burckhardt, Staffelführer Jagdstaffel 25.
10. Alex Imrie, *German Fighter Units 1914–May 1917*, p.30, London 1978.

8. Factory Ffinish of the First Halberstadt and Albatros Fighters

8.1 The Colors of the Factory Finish of the Halberstadt Fighters

1. Hans-Joachim Buddecke, *El Schahin – Der Jagdfalke*, p.94–95, Berlin 1918.
2. Hans-Joachim Buddecke, *El Schahin – Der Jagdfalke*, p.103, Berlin 1918.

8.2. Factory Finishes of the Fuselages of the Albatros D I, D II, and D III

1. Flug-Meldebuch der AOK der 6. Armee, Bund 318, Bayerisches Hauptstaatsarchiv, Abt. 4, München.
2. Alex Imrie, *German Fighter Units 1914–May 1917*, p.27 and 31.

Sources
Publications

Arbeitsgemeinschaft Luftfahrt 1900 – 1920, *Das Propellerblatt*, 1/2001.

Arbeitsgemeinschaft Luftfahrt 1900 – 1920, *Das Propellerblatt*, 2/2001.

Arbeitsgemeinschaft Luftfahrt 1900 – 1920, *Das Propellerblatt*, 8/2003.

Arbeitsgemeinschaft Luftfahrt 1900 – 1920, *Das Propellerblatt*, 18/2007.

Arbeitsgemeinschaft Luftfahrt 1900 – 1920, *Das Propellerblatt*, Nr. 38.

Bronnenkant, Lance, *Oswald Boelcke, the Red Barons Hero*, Aeronaut Books, USA 2018.

Bronnenkant, Lance, *The Blue Max Airmen*, Vol. 1, Aeronaut Books, USA 2012.

Bronnenkant, Lance, *The Blue Max Airmen*, Vol. 2, Aeronaut Books, USA 2012.

Bronnenkant, Lance, *The Blue Max Airmen*, Vol. 3, Aeronaut Books, USA 2013.

Bronnenkant, Lance, *The Blue Max Airmen*, Vol. 4, Aeronaut Books, USA 2013.

Bronnenkant, Lance, *The Blue Max Airmen*, Vol. 5, Aeronaut Books, USA 2014.

Bronnenkant, Lance, *The Blue Max Airmen*, Vol. 7, Aeronaut Books, USA 2015.

Bronnenkant, Lance, *The Blue Max Airmen*, Vol. 11, Aeronaut Books, USA 2018.

Buddecke, Hans-Joachim, *El-Schahin – Der Jagdfalke*, Berlin 1918.

Chef des Generalstabes des Feldheeres, Weisung für den Einsatz und die Verwendung von Fliegerverbänden innerhalb einer Armee, Mai 1917.

Chef des Generalstabes des Feldheeres, Weisung über den Einsatz von Jagdstaffeln, 25.10.1917.

Ebenhardt von, Walter, *Unsere Luftstreitkräfte*, semi-official casualty list of the German Army Air Service 1914–1918, Berlin 1930.

Ferko, Ed, *Fliegertruppe 191 –1918* Volume 1, Ohio 1980.

Grosz, Peter M., *Halberstadt Fighters*, Berkhamsted 1996.

Grosz, Peter M., Windsock Datafile 100, Berkhamsted 2003.

Gastreich, Steffen, Waiss, Walter, *Jagdstaffel Boelcke 1914–1918*, Aachen 2016.

Goote, Thor, *Rangehen ist alles*, Berlin 1938.

Herris, Jack, *AEG Aircraft of WWI*, USA 2015.

Herris, Jack, *Albatros Aircraft of WWI* Volume 1, Early Two-Seaters, USA 2016.

Herris, Jack, *Albatros Aircraft of WWI* Volume 4, Fighters, USA 2017.

Herris, Jack, *DFW Aircraft of WWI*, USA 2017.

Herris, Jack, Scott, Josef, *Fokker Aircraft of WWI*, Volume 2 Eindeckers, USA 2021.

Herris, Jack, Leckscheid, Jörn, *Fokker Aircraft of WWI* Volume 3 Early Biplane Fighters, USA 2021.

Herris, Jack, *Halberstadt Aircraft of WWI* Volume 2, USA 2020.

Herris, Jack, *LVG Aircraft of WWI* Volume 1, B-Types & C I, USA 2019.

Imrie, Alex, *Pictorial History of the German Army Aur Service 1914–1918*, London 1971.

Imrie, Alex, *German Fighter Units 1914–May 1917*, London 1978.

Imrie, Alex, *Pictorial History of the German Army Air Service 1914–1918*, London 1971.

Hoeppner von, Ernst, *Deutschlands Krieg in der Luft*, Leipzig 1921.

Knötel d.J., Herbert, Pietsch, Paul, Baron Collas, *Das deutsche Heer, Friedensuniformen bei Ausbruch des Weltkrieges*, Band I, II and III, Stuttgart 1982.

Kriegswissenschaftliche Abteilung der Luftwaffe, *Mobilmachung, Aufmarsch und erster Einsatz der deutschen Luftstreitkräfte im August 1914*, Berlin 1939.

Méchin, David, *The WWI French Aces Encyclopedia*, Volume 4, Aeronaut Books 2021.

Napp, Niklas, *Die deutschen Luftstreitkräfte im 1. Weltkrieg*, Paderborn 2017.

Neumann, Georg-Paul, *Die deutschen Luftstreitkräfte im Weltkrieg*, Berlin 1920.

Nowarra, Heinz J., *Eisernes Kreuz und Balkenkreuz*, Mainz 1968.

O'Connor, Neal, *Aviation Awards of Imperial Germany in World War I*, Vol. IV.

Over the Front, Volume 3, No. 4, USA 1988.

Potempa, Harald, *Die königlich bayerische Fliegertruppe 1914–1918* (Europäische Hochschulschriften, Reihe II, Bd. 727, Frankfurt 1977).

Richthofen von, Manfred, *Der rote Kampfflieger*, Berlin – Wien 1917.

Ryheul, Johan, *KEKs and Fokkerstaffels*, Croydon 2014.

Schmäling, Bruno & Bock, Winfried *Royal Prussian Jagdstaffel 30*, Aeronaut Books   USA 2014.

Schmäling, Bruno & Bock, Winfried, *Royal Bavarian Jagdstaffel 23*, Aeronaut Books USA 2018.

Schmäling, Bruno, *Jasta Colors* Volume 1, Aeronaut Books, USA 2020.

von Schoenermarck Alexis, (Hrsg.): *Helden-Gedenkmappe des deutschen Adels*, Verlag Wilhelm Petri, Stuttgart 1921.

Schmidt, Arthur, *Die grauen Felduniformen der Deutschen Armee*, Reprint Hamburg 1983/84.

Schnitzler, Erich, *Carl Allmenröder der bergische Flieger*, Wald 1927.

Schäfer Carl-Emil, *Vom Jäger zum Flieger*, Berlin 1918.

Sörnsen, Nils (alias Holler, Carl), *Als Sängerflieger im Weltkrieg*, Hamburg 1933.

Stacy, Michael, Hans Müller ein erstklassiger Eindeckerflieger, *Cross & Cockade* Vol. 24, No. 2, Summer 1983.

Täger, Hannes, The lost Honour of Ernst Freiherr von Althaus, *Over the Front*, Volume 29, Number 2, Summer 2014.

Tutschek, Ritter von, Adolf, Goote, *Thor (Herausgeber), In Trichtern und Wolken*, Braunschweig 1934.

Tutschek, Ritter von, Adolf, *Stürme und Luftsiege*, Berlin 1918.

Übersicht der Behörden und Truppen in der Kriegsformation, Teil 10 Luftstreitkräfte – Abschnitt B: Fliegerformationen, Berlin 1919.

Werner Prof. Dr., Johannes, *Boelcke*, Leipzig 1932.

Werner Prof. Dr., *Briefe eines deutschen Kampffliegers an ein junges Mädchen*, Leipzig 1930.

VanWyngarden, Greg, *Early German Aces of World War I*, Oxford 2006.

Online Sources

www.frontflieger.de

https://de.wikipedia.org/wiki/Hanns_Braun_(Leichtathlet).

https://www.hall-of-fame-sport.de/mitglieder/detail/Hanns-Braun.

Archival Records
Bayerisches Hauptstaatsarchiv, Abt 4, München

Akte Iluft 80, Könglich Bayerische Inspektion des Militär-, Luft- u. Kraftfahrwesens, Akte MMJO V K 14 /10.

A.O.K. 6 Flieger Nachrichten- und Verfolgungsstelle.
Armee-Tagesbefehl 21.09.1916, Stabsoffizier der Flieger 6, wöchentlicher Tätigkeitsbericht Stofl 6 vom
    25.09.1916.
Bund 35 Akte 1.6. Landwehr Division.
Flug-Meldebuch der AOK der 6. Armee, Bund 318.
Kriegstagebuch des XV. bayerischen RK., Bd. 578.
KA ILUFT204 - Flugzeuge Armeeflugpark 6, La Briquette, 31.10.1916.
Personal records:
    Diekmannshemke, Gustav, Kriegsstammrolle 17961/35; 17962/19; 18153/52.
    Grünzweig, Fritz, OP 20214, Kriegsstammrolle 18005/8.
    Müller Ritter von, Max, Akte MMJO V K 14 / 10.
    Pfleiderer, Georg, Personalakte OP 9301, Kriegsstammrolle 17968/158.
Rangliste des Armee-Flugparks 6.
Tägliche Meldungen der Flieger-Nachrichten und Verfolgungsstelle.
Wochenbericht des Stabsoffiziers der Flieger, 6.Armee vom 10.10.1916.
Übersicht der Behörden und Truppen in der Kriegsformation, Teil 10 Luftstreitkräfte – Abschnitt B:
    Fliegerformationen.

Bundesarchiv Freiburg

Personalakte 6/3092 Raimund Armbrecht.

Rainer Absmeier Collection

Various photos.
Documents Jasta 1.

Winfried Bock Colletion

German Pilots with Four or More Aerial Victories, unpublished manuscript.
Combat report of British squadrons, summarized in the daily intelligence summary of the RFC/RAF "War
    dairy" 1916 – 1918.
Kaus, Erich, Photos, documents, correspondence Dr. Bock and Erich Kaus.
War diary excerpt Jagdstaffel 1, by Erich Tornuß ca. 1930, revised and supplemented by Dr. Gustav Bock and
    Winfried Bock.
War diary excerpt Jagdstaffel 2, *ibid.*
War diary excerpt Jagdstaffel 3, *ibid.*
War diary excerpt Jagdstaffel 4, *ibid.*
War diary excerpt Jagdstaffel 5, *ibid.*
War diary excerpt Jagdstaffel 6, *ibid.*
War diary excerpt Jagdstaffel 7, *ibid.*
War diary excerpt Jagdstaffel 8, *ibid.*
War diary excerpt Jagdstaffel 9, *ibid.*
War diary excerpt Jagdstaffel 10, *ibid.*
Sippel, Hans, KEK-Ensisheim, Jasta 16, photos, notes, correspondence with Dr. Gustav Bock.
Ulmer, Alfred, Jasta 8, Photos, notes.
Tornuß, Erich, handwritten transcriptions about the painting of the aircraft of various Jagdstaffeln, based on
    his notes from the war diaries and combat reports in the Reichsarchiv ca. 1930.

Lance Bronnenkant Collection

Various photos.
Short biography about Friedrich Karl von Preußen.

Jack Herris Collection

Varous photos.

Alex Imrie Collection

War diary excerpt Jagdstaffel 1.
Private collection, of the following airmen:
    Budde von, Hans-Hermann, Jasta 29, 15, photos, documents, handwritten notes on the conversation.
    Burckhardt, Karl-Friedrich, Jasta 25, photos, documents, handwritten notes on the conversation.
    von Hippel, Hans-Joachim, photos, notes.
    Janzen, Johann, Kagohl 2, Kasta 12, Jasta 23, Staffelführer Jasta 6, photos, documents, handwritten notes on the conversation.
    Lenz, Alfred, Jasta 4, 14, Staffelführer Jasta 22, photos, documents, handwritten notes on the conversation.
    Ray, Franz, Jasta 1, 28, Staffelführer Jasta 49, photos, documents, handwritten notes on the conversation.
    Student, Kurt, General, photos, documents, handwritten notes on the conversation.

Reinhard Kastner Collection

Kampfgeschwader der Obersten Heeresleitung III, Document on the markings of the Kampfstaffeln 13 – 18.
War diary compilation of Bavarian Feldflieger-Abteilung 6.
War diary compilation of Bavarian Feldflieger-Abteilung 7.
War diary compilation of Bavarian Feldflieger-Abteilung 8.
War diary compilation of Bavarian Feldflieger-Abteilung 9.
War diary compilation of Bavarian Fliegerabteilung 304.
Walz, Franz, Brieftauben-Abteilung-Ostende, Kagohl I, Staffelführer Jasta Boelcke, Jasta 19 and 34, Fliegerabteilung 304, photos, documents, notes.

Thorsten Pietsch Collection, www.frontflieger.de

Information about German pilots of World War I.

Terry Phillips Collection

Kampfgeschwader III, photos of Kampfstaffeln 16 und 17.

Jörn Leckscheid Collection

Various photos.
Unpublished study of the various Albatros D I and D II types and their manufacturers.

Colin Owers Collection

Various photos.

Bruno Schmäling Collection

Various letters from Fokker monoplane pilots to Anthony Fokker 1915–1916, unpublished.
Geschwaderflüge, type-written manuscript of Manfred von Richthofen.
Lersner von Dr., Albert, letter to Bruno Schmäling from 21.05.1982.
    Privat collection of the following airmen:
Allmenröder, Carl, FFA 18, Jasta 11, photo-album, documents, notes.
Althaus von, Ernst, photo-album, documents, notes.

Armbrecht, Raimund, Jagdstaffel 1, photo-album, notes.

Arntzen, Heinz, Jasta 15, Staffelführer Jasta 50, photo-album, personal diary, documents, notes, correspondence, audio tape interview, handwritten notes on the conversation, September 1978.

Auer, Hans, Feldflieger- Abteilung 9b, Jasta 26, photo-album, documents, notes.

Böning, Walter, Jagdstaffel FFA 6b, Jasta 19, 76, photo-album, war-time letters, documents, notes, handwritten notes on the conversation.

Boes, Hans, Kasta 36, Jagdstaffel 34, photo-album, documents, notes.

Brettel, Hermann, Jagdstaffel 10, photo-album, documents, notes.

Bülow von, Walter, Fliegerabteilung 300 „Pascha", photo-album, documents, notes.

Christensen, Waldemar, Kagohl 1, Jasta 5 und 46, photos, notes.

Crailsheim von, Hans-Jürgen, Information about Kurt Freiherr von Crailsheim.

Eckenberg, Wilhelm, photo-album, documents, notes.

Fuchs, Otto, FA (A) 292, Jasta 30 und 77, Staffelführer Jasta 35, photo-album, documents, notes, handwritten notes on the conversation.

Frommherz, Hermann, Kagohl IV, photo-album, documents, notes.

Geigl, Heinrich, Kasta 36, Jagdstaffel 34 und 16, photo-album, documents, notes.

Gusnar von, Ernst, Feldflieger-Abteilung 62, photo-album, documents, notes, personal diary.

Heldmann, Alois, FA 57, 59, Jasta 10, handwritten notes on the conversation.

Jacobs, Josef, FFA 11, Fokkerstaffel West, Jasta 22, Staffelführer Jasta 7, photo-album, documents, notes, personal diary, handwritten notes on the conversation.

Jacobsen, Fritz John, FA 1, Jasta 31 und 73, photo-album, documents, notes, type-written report about his time with Jasta 31, handwritten notes on the conversation.

Jensen, Johann, Jasta 57, photo-album, documents, notes.

Junginger, Hans, mechanic, FFA 62, KEK II and III, Jagdstaffel 10, photo-album, documents, notes.

Junk Gen. a. D., Führer A.O.K. Staffel der 3. Armee, Staffelführer Jasta 9, handwritten notes on the conversation.

Kempf, Fritz, Kagohl IV, Jasta Boelcke, photo-album, documents, notes.

Leinemüller, Xaver mechanic, Jasta 10, photo-album, documents, notes, handwritten notes on the conversation

Lersner von, Rolf, Kagohl I, Jagdstaffel Boelcke, photo-album, documents, notes

Mai, Josef, photos, flightlog, personal diary, handwritten notes on the conversation

Mannschott, Friedrich, Jasta 7, photo-album, documents, notes.

Mohr, Hans, mechanic Jasta 9, photo-album, documents, notes

Nebel, Rudolf, Jasta 5, Staffelführerer Ke.St. 1a, 1b and Jasta 90, photo-album, war-flight-log, documents, notes, handwritten notes on the conversation.

Parschau, Otto, letter from 25.05.1915 to Anthony Fokker.

Reiss, Eckehard, mechanic, AFA 211 und Fokkerstaffel Falkenhausen, photo-album, documents, notes.

Rumple, Theodor, FA (A) 280, Jasta 26, 16 und 23, photo-album, war-flightlog, documents, notes, letters, handwritten notes on the conversation.

Schäfer, Emil, photo-album, documents, notes.

Schobinger, Victor, Jasta 12, photo-album, documents, notes, letters, handwritten notes on the conversation.

Schoenebeck von, Karl-August, Jasta 11, 59, Staffelführer Jasta 33, photo-album, documents, notes, handwritten notes on the conversation.

Schoenfelder, Kurt, Jagdstaffel 7, photo-album, notes.

Student, Gen. a.D., Kurt, photo-album, documents, notes, letters, handwritten notes on the conversation.

Thuy, Emil, FFA 53, Jagdstaffel 21 und 28, photo-album, documents, notes, typewritten reports.

Wenig, Erwin, FFA 8b, Kampfeinsitzer-Kommando Ensisheim, Kampfeinsitzer-Kommando der Obersten Heeresleitung, Jagsta 16, 28 und 80, photo-album, documents, notes, letters.

Zilcher Ferdinand, Jasta 1, handwritten notes on the conversation.

Michael Schmeelke Collection

Pfeiffer, Hermann, photo-album, documents, notes.

Herbert Schulz Collection

Handwritten notes from the interviews with former pilots of the Jagdstaffel Boelcke from the 1920s.
Compilation of the markings of the aircraft of Jagdstaffel 2 (Boelcke), 1916, from the 1920s.

Maton Szigeti Collection

Pier, Wilhelm, Feld-Flieger-Abteilung 23, photo-album, notes.

Hannes Täger Collection

Baldamus, Hartmut, photo-album, documents, notes.
Research on Erich Koehler by Hannes Täger.

Greg VanWyngarden Collection

Flieger-Ersatz-Abteilung 10, contemporary color painting.
Goettsch, Walter, Jagdstaffel 8, photos, research on the painting of the aircraft 1916.
Holler, Carl, Jasta 6, photo-album, notes.
Keudell vonn, Hans, Jagdstaffel 1, photo-album, notes.
Kampfgeschwader II, Kampfstaffel 11, photos, notes.
L'Aerophile, Frankreich, 1916, Excerpts from transcripts.
Colour drawing of the DFW B.I B.451/14 captured by the French.
Color drawing of an Aviatik B in 1915 at Flieger-Ersatz-Abteilung 10 in Böblingen.

Tobias Weber Collection

Buddecke, Hans-Joachim, photo-album, documents, notes.
Grünzweig, Fritz, KEK-Ensisheim, Jasta 16, documents, sketches, and drawings.

Reinhard Zankl Collection

Various photos.
Feldflieger-Abteilung 34, contemporary post-card with handwritten notes.
Jagdstaffel 3, photos, handwritten notes.
Compilation of D-aircraft delivered to the front from September 1916 to January 1917 based on documents in Bavaria.

Printed in Great Britain
by Amazon